Birkhäuser

Advanced Courses in Mathematics
CRM Barcelona

Centre de Recerca Matemàtica

More information about this series at http://www.springer.com/series/5038

Paul Glendinning • Mike R. Jeffrey

An Introduction to Piecewise Smooth Dynamics

Editors for this volume:
Elena Bossolini, Technical University of Denmark
J. Tomás Lázaro, Universitat Politècnica de Catalunya
Josep M. Olm, Universitat Politècnica de Catalunya

Paul Glendinning
School of Mathematics
University of Manchester
Manchester, UK

Mike R. Jeffrey
Department of Engineering Mathematics
University of Bristol
Bristol, UK

ISSN 2297-0304 ISSN 2297-0312 (electronic)
Advanced Courses in Mathematics - CRM Barcelona
ISBN 978-3-030-23688-5 ISBN 978-3-030-23689-2 (eBook)
https://doi.org/10.1007/978-3-030-23689-2

Mathematics Subject Classification (2010): 34A36, 34C23, 37E05, 37G10, 37E35, 37G35

This book is published under the imprint Birkhäuser, www.birkhauser-science.com by the registered company Springer Nature Switzerland AG
The registered company address is: Gewerbestrasse 11, 6330 Cham, Switzerland

Foreword

The study of piecewise-smooth dynamics encompasses the theory and applications of how dynamical systems (in general, smooth almost everywhere) behave when they undergo discontinuities at isolated thresholds. In this sense, the main aim of the *Intensive Research Programme* on *Advances in Nonsmooth Dynamics* held at the Centre de Recerca Matemàtica (Bellaterra, Spain), from February to April 2016, was to take stock of the recent progress in the area and to discuss on the current open challenges and future research topics.

Along twelve definitely intense weeks the programme attracted more than seventy participants from all over the world, including senior and junior researchers, postdocs, and PhD students. A group of permanent residents was complemented with temporary visitors whose research interests gathered according to a weekly distribution of thematic subjects. Four scientific events were organized within the programme: an opening conference on "Open Problems in Nonsmooth Dynamics", a workshop on "Climate Modeling", an advanced course on "Piecewise-Smooth Dynamical Systems", and a summary session on "Nonsmooth Dynamics, the Way Forward".

This volume of the series *Advanced Courses in Mathematics CRM Barcelona* gathers the lectures delivered at the above-mentioned advanced course. It took place during the week from 11th to 15th April, 2016. The first chapter, authored by Mike R. Jeffrey (University of Bristol, UK), is an introduction to the dynamics of piecewise-smooth flows, while the second chapter, by Paul Glendinning (University of Manchester, UK), deals with piecewise-smooth maps.

We close these lines thanking the authors, Mike Jeffrey and Paul Glendinning, for this elaborated version of the notes, the attendees for their active participation and feedback, and the institutions which provided specific support for the organization of this course, namely, the Engineering and Physical Sciences Research Council (EPSRC), the Societat Catalana de Matemàtiques (SCM), the company Maths for More, and the Institute for Mathematics and its Applications (IMA). Finally, our special thanks to the Centre de Recerca Matemàtica (CRM) for the warm hosting and great support provided not only in the course, but also during all the *Intensive Research Programme* on *Advances in Nonsmooth Dynamics*.

Lyngby, Barcelona, April 2017

Elena Bossolini
Tomás Lázaro
Josep M. Olm

Contents

Introduction

The study of nonsmooth dynamics has a long history, but has been recently enlivened by the introduction of new techniques, the discovery of new phenomena, and the explosion of new practical disciplines applying nonsmooth dynamical modeling. The more formal name for our field is *piecewise-smooth* dynamics, concerning the theory and applications of how dynamical systems that are smooth almost everywhere behave when affected by discontinuities at isolated thresholds.

This volume presents an informal course at graduate level, aimed at mathematicians, scientists, and engineers, studying models that involve a discontinuity, or studying the theory of nonsmooth systems for its own sake. These notes are derived from lectures given at the Advanced School on *Piecewise-Smooth Dynamical Systems*, organized at the Centre de Recerca Matemàtica (Bellaterra, Catalonia), during April 11–15, 2016. The School was organized with support from the Engineering and Physical Sciences Research Council (EPSRC), the Societat Catalana de Matemàtiques (SCM), WIRIS, and the Institute for Mathematics and its Applications (IMA). The course comes in two parts: flows and maps (or continuous and discrete time systems). Each part introduces the applications and main theoretical techniques, of piecewise-smooth dynamics, some long-established and others newly emerging.

An introduction to the dynamics of *piecewise-smooth flows* is authored by Mike Jeffrey. The chapter starts with a few key points in the history and applications of discontinuities from mechanics and control systems, and some of the key figures in setting up a general theory for differential equations with 'discontinuous right-hand sides'. The methods of inclusions versus combinations are discussed, after which we introduce the elementary dynamics of crossing and sliding at a discontinuity surface. The analytical methods of switching layers and layer variables are presented for one switch and then for multiple switches. The concepts of discontinuity-induced phenomena and determinacy breaking are introduced, along with the definitions of stability, equivalence, and bifurcation in piecewise-smooth flows. We end by introducing some of the novel attractors and bifurcations that offer a hint at what remains to be discovered, including the present state of zoology of singularities in the plane.

An introduction to the dynamics of *piecewise-smooth maps* is authored by Paul Glendinning. After a discussion of why piecewise-smooth maps are interesting, the course moves into their phenomenology, and reviews some techniques from smooth theory. The main topics are then: basic stability analysis, piecewise monotonic maps of the interval, rotation-like maps, gluing bifurcations (aka Big Bang bifurcations and period-adding), an introduction to renormalization, decomposition theorems and a brief guide to kneading theory, piecewise-smooth maps of the plane including the Lozi and border collision normal forms, piecewise isometries, bounding regions, periodic orbits and resonance, robust chaos, and two-dimensional attractors. The course concludes with a discussion of challenges in higher dimensions, particularly concerning periodic orbits, N-dimensional attractors, and analogies with smooth cases.

The Barcelona course itself included a series of guest lectures, introducing course participants to some of the current research topics in applications of piecewise-smooth dynamics. The guest lecturers were researchers in residence at the Intensive Research Program on Advances in Nonsmooth Dynamics at the CRM, February 1 – April 29, 2016. A flavour of these contributions can be found in the Extended Abstracts of the Intensive Research Program [15].

Chapter 1

Piecewise-smooth Flows

1.1 Introduction

This course is about the geometry of piecewise-smooth dynamical systems. The solutions of a system of ordinary differential equations, such as

$$\dot{\boldsymbol{x}} = \boldsymbol{f}(\boldsymbol{x}), \tag{1.1}$$

where $\boldsymbol{x} = (x_1, x_2, \ldots, x_n)$ is some n-dimensional vector or variable, and $\boldsymbol{f}$ is an n-dimensional vector field, can be pictured as trajectories (or *orbits*) in space (for example, $\mathbb{R}^n$ or some subset of it). Those trajectories are organized by various singularities, separatrices, and invariant sets, whose geometry can be studied in great generality. A loss of continuity in the differential equations greatly adds to the richness of that geometry.

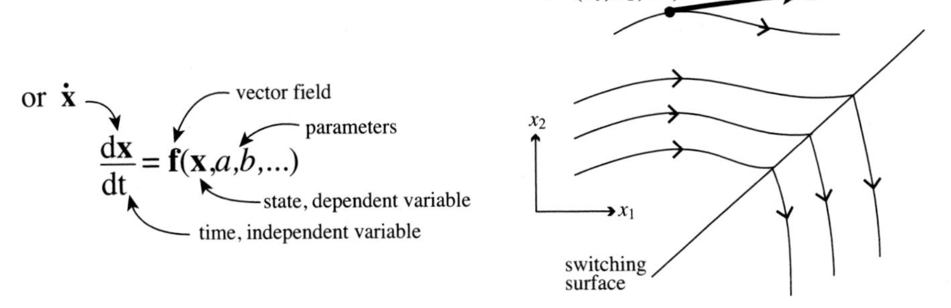

Figure 1.1: The vector field $\boldsymbol{f}$ tells us the velocity with which some flow evolves through a state $\boldsymbol{x}$. If the flow is non-differentiable, then $\boldsymbol{f}$ is discontinuous.

Piecewise-smooth equations are smooth except at isolated thresholds called *switching surfaces*. Solutions of those equations are continuous, but they may 'kink' at a switching surface, becoming non-differentiable, and possibly non-unique.

P. Glendinning, M. R. Jeffrey, *An Introduction to Piecewise Smooth Dynamics*, Advanced Courses in Mathematics - CRM Barcelona, https://doi.org/10.1007/978-3-030-23689-2_1

The mathematician Alexei Fedorovich Filippov set out methods for solving piecewise-smooth differential equations, and these have been adopted as standard. Recently, however, with the discovery of new singularities and singular phenomena, we have understood that more is needed. This series of lectures will provide readers with the tools to delve more deeply into the world of nonsmooth dynamics.

We will focus on:

- geometry of piecewise-smooth vector fields,
- general methods for solving and analysing them,
- their key notions of stability and bifurcations.

Some important current topics that we will not cover, but you may wish to look up, include:

- special cases: e.g., piecewise linear or continuous non-differentiable vector fields, hybrid/impact systems;
- modeling non-idealities: e.g., smoothing, noise, delay, etc., and the various types of 'regularization', a word that comes up a lot as an open problem in current piecewise-smooth systems.

You may also want to look into simulation methods: piecewise-smooth systems require special consideration when simulating. There are event detection routines built into Matlab and Mathematica. There are ways of making continuation tools like AUTO or MatCont work with discontinuities (often by smoothing them out), including the AUTO-derived TcHat. Filippov's solving methods (we'll discuss these in the course) have even been built into tools like Mathematica. But these are not up-to-date in the many advances in theory that we have seen in the last decade. Use them all with care and critical judgement.

1.1.1 Nonsmooth dynamics in a nutshell

You will notice that this first chapter on piecewise-smooth flows devotes most of its content to considering how we obtain a well-defined flow from a discontinuous vector field, leaving only limited space to begin introducing the fascinating phenomena that result. This is because so much is possible when a flow passes through a discontinuity, that our concept of solutions must be of sufficient detail and sufficient generality to explore those possibilities. This is in contrast to the second chapter on maps, where the vanishing likelihood of encountering the discontinuity itself, when evolving in discrete steps, simplifies the field considerably, the solution concept is easily set up, and the study can proceed directly to studying the dynamics.

When we have introduced the solution concept for piecewise-smooth flows properly, we will see that there are really just a few basic elements needed to begin building up an understanding. First the vector field.

The vector field is discontinuous at the switching surface. A *sliding* vector field may be induced on the surface, and we shall see how to define this.

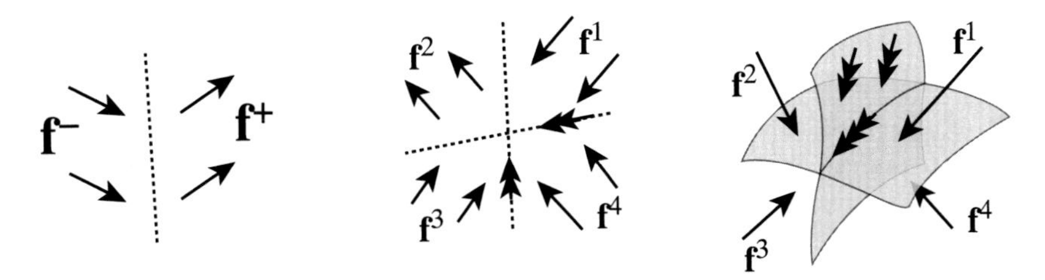

Figure 1.2: Discontinuities along one threshold in two dimensions, two thresholds in two dimenions, and two thresholds in three dimensions.

Locally, solutions take certain simple forms. Away from the switching surface they will be smooth unique curves, thanks to the existence and uniqueness of solutions of differentiable dynamical systems. At the surface, however, they might cross through the discontinuity, or they might slide along it, following a sliding vector field.

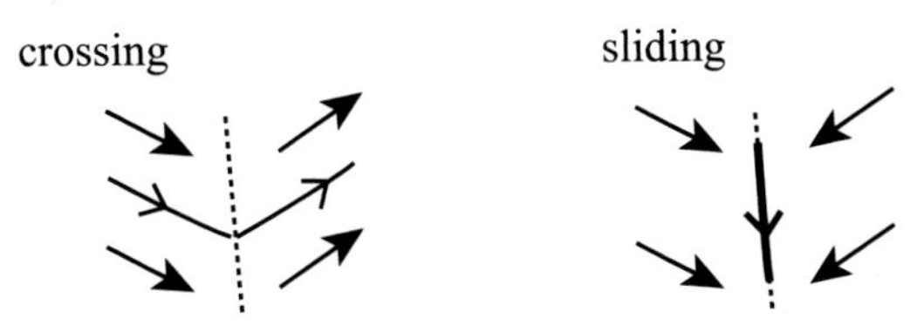

Figure 1.3: Typical trajectories through or along a switching surface.

The switching surface has a lower dimension than the surrounding space, so if sliding occurs, the space the solutions occupy changes, see fig. 1.4. This results in non-uniqueness. When sliding is:

- attractive, solutions stick to the switching surface, and then many solutions will all evolve onto the same trajectory in forward time. The history of any point (shown in the left part of fig. 1.4) in attractive sliding is therefore non-unique — this is common in physics as mechanical 'sticking'.
- repulsive, solutions escape a switching surface, and then the sliding mode has many possible future trajectories (shown in the right part of fig. 1.4). The future of any point in repulsive sliding is therefore non-unique — determinacy is broken.

The only other things we *must* add to these are the elementary singularities. In differentiable vector fields, the commonly encountered singularity is a steady state or 'equilibrium', where $\boldsymbol{f} = 0$ in a system $\dot{\boldsymbol{x}} = \boldsymbol{f}$. At a discontinuity we encounter a new kind of steady state, a *sliding equilibrium*, fig. 1.5, (called a

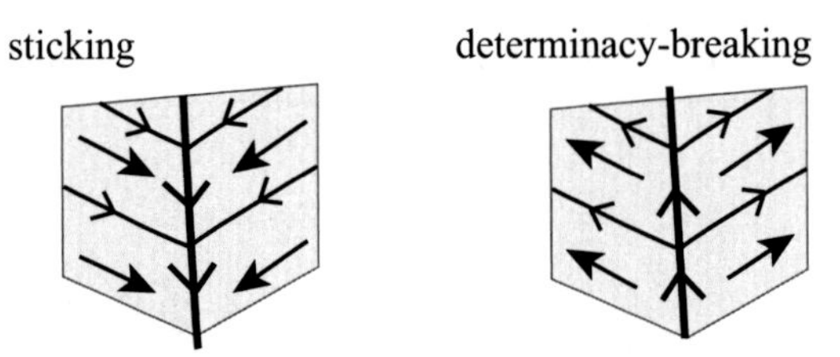

Figure 1.4: Sliding along the switching surface leads to sticking (if attractive) or determinacy-breaking (if repulsive).

'pseudo'-equilibrium in many texts, but they are genuine steady states, as we shall see, so using the term 'pseudo' is misleading).

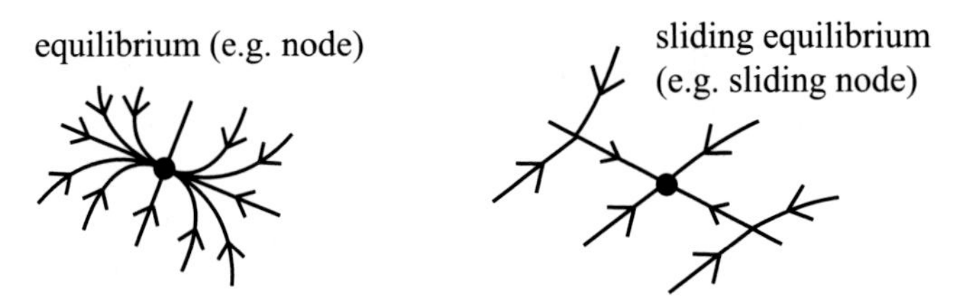

Figure 1.5: The basic steady-state singularities.

In piecewise-smooth systems there is much more to local dynamics than studying equilibria. It is often the transient (i.e., non-stationary) dynamics that creates the most interesting effects. In addition we must consider stationarity relative to the switching surface, i.e., *tangency* of solutions to the switching surface.

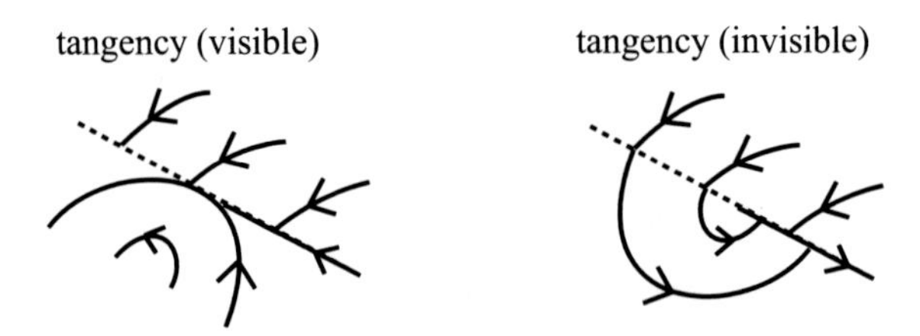

Figure 1.6: The basic transient singularities, characterizing stationarity between the flows and the switching surface.

Keeping these few elements in mind will be of great assistance in gaining some intuition for the (at first strange) terrain of what we have come to call informally: *Nonsmoothland.*

1.2 History and applications

Dynamical theory for smoothly evolving systems came a long way in the last century. The most interesting phenomena that are now familiar in dynamical systems occur because of nonlinearity. Nonlinearity means the rate of change of a system is itself varying as we move around the system, so any local approximation will not represent the wider system well. Most important for us is what this means in terms of dynamical behaviour: if we change some variable or parameter, then the system's behaviour may not respond proportionally (unlike a linear system), and we obtain phenomena like bifurcations, chaos and complexity.

Continuity and differentiability of such equations have been vital to taming the complications of nonlinearity. Systems which are *not* differentiable or are *not* continuous, however, have been studied as long as there have been dynamical systems. Collisions between rigid bodies result in a discontinuous jump in contact force. Electrical switches discontinuously turn on/off the current in a circuit. People in societies discontinuously switch from following one rule or trend to another. The intervals of smooth evolution before and after a switch are well described by standard (nonlinear) dynamical theory. To stitch those 'before' and 'after' systems together in more than an ad hoc fashion requires a general set of tool to understand the effect of discontinuities, via the theory of piecewise-smooth dynamics.

Discontinuities are often involved when different objects or subsystems interact. At a point where the equations are not differentiable, or not even continuous, almost nothing from standard 'smooth' dynamical systems theory can be applied.

1.2.1 Ancient history

One of the most familiar physical forces to us is also one of the most complex, and a prime example of how discontinuities complicate one of the most fundamental processes of interaction: friction.

The resistance force of friction between two rigid bodies in contact with rough unlubricated contact surfaces has a long and contentious history. It goes back to the greek philosophers, but let us start with the seeds of the modern theory:

- in 1500 Leonardo da Vinci shows that frictional resistance depends on load, but not on contact area;
- in 1699 Amontons rediscovers da Vinci's laws, and describes friction as the work done to overcome — through wear and deformation — the surface roughness between two objects;
- in 1700 Desagulier shows that friction does *not* depend on surface roughness, seeming to contradict Amonton's theory;
- in 1750 Euler shows that static friction force is greater than kinetic friction force, so more force is required to instigate motion of a static object, than is required to keep an object in motion;

- in 1785 Coulomb added clarity and depth to the former theories, especially those of da Vinci and Amontons;
- in 1950 Bowden and Tabor clarified the role of contact area and surface roughness, showing that friction does depend on the *true* contact area, which is often less than the apparent contact area (i.e., the full contact surface);
- 1950 and into the present, the field of Tribology remains active, studying speed-dependence and friction memory, the different friction characteristics at different scales of length, time, or speed, and remains important, for example, to the health, automotive, and energy industries.

This is a very fleeting description, meant only to highlight the complexity of a contact force that we use and interact with every moment of every day. The most enduring part of this story — that friction depends principally on load only — is encapsulated in the now commonly adopted law

$$m\ddot{x} = -\mu F_N \tag{1.2}$$

where m is the mass of an object with displacement x, moving at speed $\dot{x}$, on a surface moving at speed v, creating a normal reaction force F_N on the object. The quantity μ is the coefficient of friction, given for some constant μ_k by

$$\mu = \mu_k \,\mathrm{sign}(\dot{x} - v) = \mu_k \times \begin{cases} +1 & \text{if } \dot{x} > v, \\ -1 & \text{if } \dot{x} < v, \end{cases} \tag{1.3}$$

where $\dot{x} > v$ means the object is slipping to the right and $\dot{x} < v$ means the object is slipping to the left (relative to the surface).

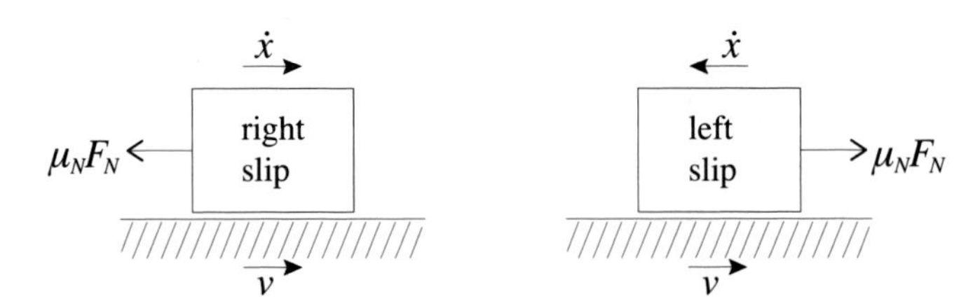

Figure 1.7: The switching force of friction.

We have in friction our first discontinuous system. It says that as an object changes from slipping right ($\dot{x} > v$) on a surface to slipping left ($\dot{x} < v$), the friction force jumps abruptly between $-\mu_k F_N$ and $+\mu_k F_N$. What happens in between? What complications does the jump introduce into the dynamics? These are the questions of piecewise-smooth dynamical systems theory.

The other very common piecewise-smooth system, typically encountered in highschool mechanics even before calculus teaches us to deal with smooth systems, is a collision. Take a block of mass m, with position x, driven by a force f, colliding with a wall at position c,

$$m\ddot{x} = f: \quad \dot{x} \mapsto -r\dot{x} \text{ if } x = c \text{ and } \dot{x} > 0, \tag{1.4}$$

where r is the coefficient of restitution. This is an *impact* system, composed of a differential equation (left part) and a discrete impact map (right part).

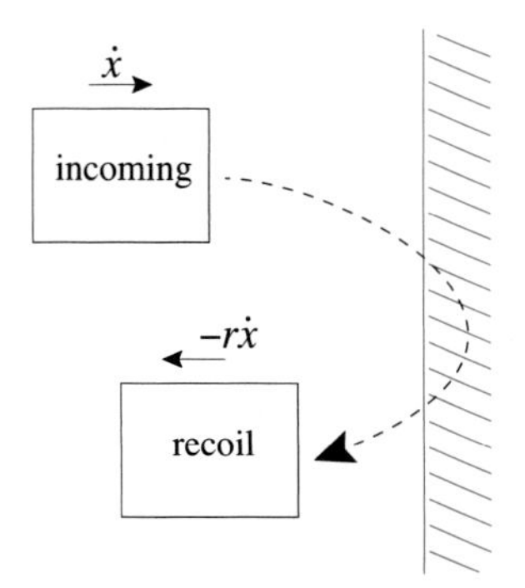

Figure 1.8: The discontinuity in velocity resulting from impact with a hard wall.

Because this mixes up continuous and discrete time evolution, it crosses the divide between the two chapters of this course, and opens the way to study of the incredibly general field of sytems that are *hybrids* of maps and flows (see, e.g., [12, 21, 99]), which we will not delve in to here (another example of a hybrid system is cellular mitosis: a cellular organism grows continuously until it triggers a discrete change in mass corresponding to mitosis; see, e.g., [27]).

However, usually a map just represents a jump through some fast continuous interval of motion. In the case above, the restitution map $\dot{x} \mapsto -r\dot{x}$ represents a jump through a continuous impact phase. Alternatively we could model this as

$$m\ddot{x} = f - k\,\text{step}(x - c), \quad \text{step}(x - c) = \begin{cases} 1 & \text{if } x > c, \\ 0 & \text{if } x < c, \end{cases} \tag{1.5}$$

introducing a large wall stiffness k. This now has continuous time solutions that are non-differentiable at $x = c$, and falls back under our topic of piecewise-smooth flows.

1.2.2 History of piecewise-smooth dynamical theory

Serious attempts to develop the mathematics of dynamical systems with discontinuities go back to the 1930s (at least). They are worth a look, because this is still a young field of research, both in theory and applications, and the ideas in these texts still strongly influence our thinking today:

- 1934 Kulebakin [68]: vibration control for aircraft DC generators;
- 1934 Nikolzky [77]: a boat rudder as a switching controller;
- 1937 Andronov, Vitt, Khaikin [5]: numerous examples are studied of mechanical engines and circuits, analyzing their dynamics and stability;
- 1953 Irmgard Fluegge-Lotz [30]: proposes discontinuous control to design missile aiming technologies;
- 1974 onwards, Vadim Utkin [95, 96, 97, 98]: develops a general design method for electronic switching ("variable structure control"), an essential step to our modern approach to solving discontinuous systems. The method was based on the work of Filippov and contemporaries, e.g., [1, 2, 28, 76], which itself didn't reach mainstream western attention until Filippov's work was translated in 1988...
- 1988 Aleksei Federovich Filippov [29]: develops the first substantial dynamical theory of "differential equations with discontinuous right-hand sides". The book actually represents a substantial Russian literature going back half a century, some references to which you will find here, others you will find in the book's substantial bibliography;
- 1990 onwards, Marco Antonio Teixeira [90, 91, 92, 93]: shows how ideas from singularity theory can be applied to study the geometry of flows near discontinuities;[1]
- The modern era: the fundamental theory (this course) and applications (to electronics, mechanics, especially power control, but also to cell mitosis, economics, predator-prey, climate, and countless more), remain active and growing fields of interest.

1.3 Inclusions and combinations

Our first task is to learn how to turn discontinuous vector fields into well-formed differential equations, and then to learn how to solve them.

1.3.1 Discontinuous vector fields

We start with a vector $\boldsymbol{x} = (x_1, x_2, \ldots, x_n) \in \mathbb{R}^n$, of the variables $x_1, x_2, \ldots, x_n$. Its time dependence is described by vector fields $\boldsymbol{f}^i$ such that

$$\dot{\boldsymbol{x}} = \boldsymbol{f}^i(\boldsymbol{x}) \quad \text{on } \boldsymbol{x} \in \mathcal{R}_i$$

[1] We have written "1990 onwards" but one of these references is earlier. Before this Teixeira phrased this work as 'divergent diagrams' or 'pairings of fields and functions' to circumvent the early skepticism towards discontinuous dynamical systems. Teixeira now leads one of the most fervent and successful communities in geometric piecewise-smooth dynamical theory.

or, in components,

$$(\dot{x}_1, \dot{x}_2, \ldots, \dot{x}_n) = \left(f_1^i, f_2^i, \ldots, f_n^i\right). \tag{1.6}$$

The index i is taken from a set of labels identified with the regions $\mathcal{R}_i$. The boundary between these regions is the switching surface Σ, such that the full space is given by $\Sigma \cup \mathcal{R}_1 \cup \mathcal{R}_2 \cup \cdots$

Example 1.3.1. For two regions we could take labels $i = 1, 2$, or $i = +, -$, so that, e.g., $\dot{\boldsymbol{x}} = \boldsymbol{f}^+$ for $\boldsymbol{x} \in \mathcal{R}_+$ and $\dot{\boldsymbol{x}} = \boldsymbol{f}^-$ for $\boldsymbol{x} \in \mathcal{R}_-$ (fig. 1.9, left). For four regions we may take labels $i = 1, 2, 3, 4$, or alternatively $i = ++, +-, -+, --$.

The figure shows a system with: two regions, two regions where the boundary between them is a corner, or four regions.

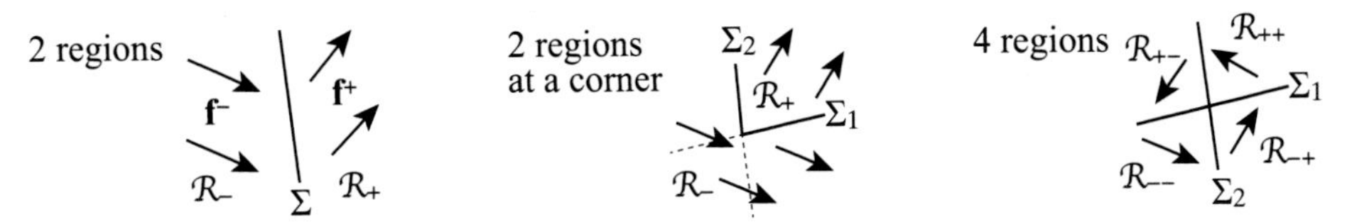

Figure 1.9: Discontinuous vector fields with $m = 2$, $m = 4$, or $m = 2$ regions; in the last example the switching surface has a corner.

In each open region $\mathcal{R}_i$ the vector fields $\boldsymbol{f}^i$ are smooth, so we can apply standard dynamical systems methodology. This course is about what then happens at the discontinuity boundary — the switching surface Σ — between them.

1.3.2 The inclusion

On the switching surface Σ the variation $\dot{\boldsymbol{x}}$ is not yet well defined. To extend $\dot{\boldsymbol{x}} = \boldsymbol{f}^i(\boldsymbol{x})$ to the boundary Σ, write

$$\dot{\boldsymbol{x}} \in \mathcal{F} \quad \text{where} \quad \boldsymbol{f}^i(\boldsymbol{x}) \in \mathcal{F} \quad \forall\, i: \quad \lim_{\boldsymbol{x}' \to \boldsymbol{x}} \boldsymbol{f}(\boldsymbol{x}') = \boldsymbol{f}^i(\boldsymbol{x})$$

where $\mathcal{F}$ is set-valued for $\boldsymbol{x} \in \Sigma$, and $\mathcal{F} = \boldsymbol{f}^i(\boldsymbol{x})$ for $\boldsymbol{x} \in \mathcal{R}_i$.

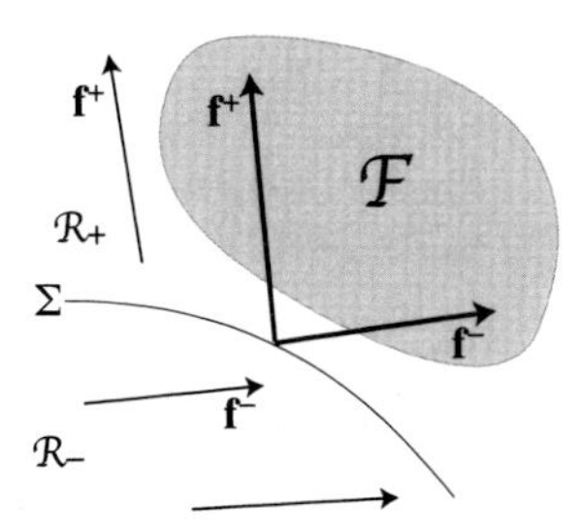

Figure 1.10: The set $\mathcal{F}$ at a switching surface Σ between vector fields $\boldsymbol{f}^+$ on region $\mathcal{R}_+$ and $\boldsymbol{f}^-$ on region $\mathcal{R}_-$.

This system is known as a *differential inclusion*, and Filippov developed their dynamical theory extensively in [29], starting with perhaps the most important result to begin with:

Theorem 1.3.2 (Existence of solutions; Filippov [29, Thm. 1, p. 77]). *Solutions of* $\dot{\boldsymbol{x}} = \mathcal{F}$ *exist if* $\mathcal{F}(\boldsymbol{x})$ *is non-empty, bounded, closed, convex, and upper semicontinuous*[2].

So if solutions to the vector field exist, forming a piecewise-smooth flow, what do they look like?

1.3.3 A classic example: Coulomb friction

Example 1.3.3. Consider a block, resting on a surface that moves at velocity v, attached to a spring of stiffness k and a damper with a coefficient c. Using the friction law we gave above, the forces on the block give us

$$\text{force} = m\ddot{x} = -c\dot{x} - kx - \lambda F_N, \quad \lambda = \begin{cases} +1 & \text{if } \dot{x} > v, \text{ “slip right”} \\ -1 & \text{if } \dot{x} < v, \text{ “slip left”} \end{cases} \tag{1.7}$$

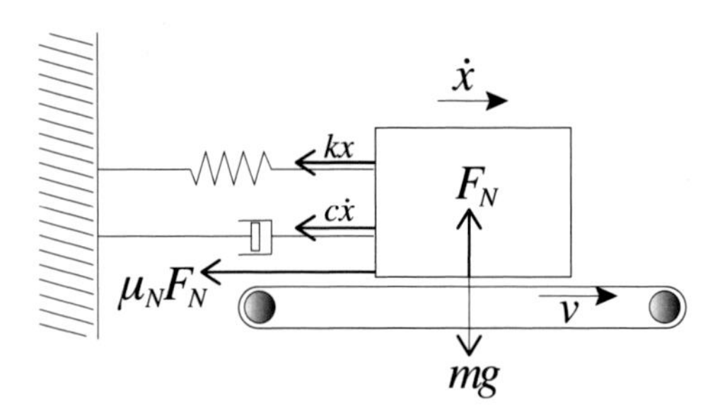

Figure 1.11: A block on a moving belt.

Written as a two-dimensional ordinary differential equation this is

$$\dot{x} = y, \qquad \dot{y} = -\tfrac{c}{m}y - \tfrac{k}{m}x - \lambda\tfrac{F_N}{m}. \tag{1.8}$$

The equilibrium at $(x, y) = (-\lambda F_N/k, 0)$ is an attractor, because (1.8) vanishes at this point, and the Jacobian matrix there,

$$\frac{\partial(\dot{x}, \dot{y})}{\partial(x, y)} = \frac{1}{m}\begin{pmatrix} 0 & m \\ -k & -c \end{pmatrix},$$

[2]Upper semicontinuity is an extension of the notion of continuity for sets. It says $\sup_{b\in\mathcal{F}(x)} \rho(b, a) \to 0$ as $p' \to p$ for $a \in \mathcal{F}(p')$ and $b \in \mathcal{F}(p)$ where $\rho(b, a) = \inf_{a\in A,\, b\in B} |a - b|$ with $|a - b|$ the distance between a and b.

has eigenvalues $(-c \pm \sqrt{c^2 - 4km})/2m$ with negative real part (giving a focus for small damping $c^2 < 4km$ and a node otherwise); these are basic results using the local stability theory of smooth dynamical systems (see, e.g., [6, 69]). Because there is only a single equilibrium, at $y = 0$, the region in which it exists (in right slip $\dot{x} > v$ or left slip $\dot{x} < v$) depends on the sign of the constant v.

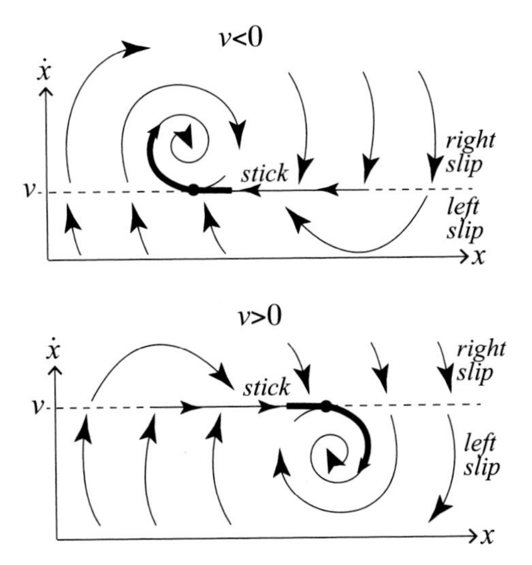

Figure 1.12: Dynamics of the oscillating block shown in the phase plane $(x, \dot{x})$.

As we trace out trajectories in this phase space, we find that the block can go from left slip to right slip or vice versa, and it can go from right or left slip to stick, then back to slip. All trajectories are eventually attracted to the equilibrium.

Sticking is represented by motion along $\dot{x} = v$, which means that the block and surface have matching speeds and are therefore stuck together.

We need a method to describe this sticking motion. This comes from solving the inclusion.

1.3.4 The switching multiplier λ and switching function σ

Take a single switch at a threshold $\sigma(\boldsymbol{x}) = 0$, in the bimodal system

$$\dot{\boldsymbol{x}} = \begin{cases} \boldsymbol{f}^+(\boldsymbol{x}) & \text{if } \sigma(\boldsymbol{x}) > 0, \\ \boldsymbol{f}^-(\boldsymbol{x}) & \text{if } \sigma(\boldsymbol{x}) < 0, \end{cases}$$

in terms of a **switching function** $\sigma: \mathbb{R}^n \mapsto \mathbb{R}$, so $\Sigma = \{\boldsymbol{x} \in \mathbb{R}^n : \sigma(\boldsymbol{x}) = 0\}$.

We can re-write this as

$$\dot{\boldsymbol{x}} = \boldsymbol{f}(\boldsymbol{x}; \lambda)$$

replacing each index i with a unique **switching multiplier** λ, such that

$$\begin{aligned} \boldsymbol{f}^+(\boldsymbol{x}) &= \boldsymbol{f}(\boldsymbol{x};+1), \\ \boldsymbol{f}^-(\boldsymbol{x}) &= \boldsymbol{f}(\boldsymbol{x};-1), \end{aligned} \quad \text{i.e., } i = \pm \;\Leftrightarrow\; \lambda = \pm 1. \tag{1.9}$$

We then define

$$\begin{aligned} \lambda &= \operatorname{sign}(\sigma(\boldsymbol{x})) && \text{for } \sigma(\boldsymbol{x}) \neq 0, \\ \lambda &\in (-1,+1) && \text{for } \sigma(\boldsymbol{x}) = 0. \end{aligned} \tag{1.10}$$

This last definition is extremely important to remember. The multiplier λ takes values ± 1 outside the switching surface, and lies inside the open interval $(-1,+1)$ on the switching surface. We shall see later how to fix the value inside this interval. We will often write for shorthand just $\lambda = \operatorname{sign}(\sigma)$, understanding the value of $\operatorname{sign}(\sigma)$ to lie in $(-1,+1)$ for $\sigma = 0$.

So the switching of the multiplier λ causes the vector field $\boldsymbol{f}(\boldsymbol{x};\lambda)$ to switch between functional forms $\boldsymbol{f}^+(\boldsymbol{x})$ and $\boldsymbol{f}^-(\boldsymbol{x})$. The dependence of $\boldsymbol{f}$ on λ tells us about *how* the jump between those modes occurs.

The function $\boldsymbol{f}(\boldsymbol{x},\lambda)$ combines $\boldsymbol{f}^+$ and $\boldsymbol{f}^-$ into a single expression where the switching is described by λ, so we may refer to $\boldsymbol{f}(\boldsymbol{x},\lambda)$ as a *combination.*

1.3.5 Combinations

The combination $\boldsymbol{f}(\boldsymbol{x};\lambda)$ is differentiable with respect to $\boldsymbol{x}$ and λ (by which we mean that the partial derivaties $\frac{\partial}{\partial \boldsymbol{x}}\boldsymbol{f}$ and $\frac{\partial}{\partial \lambda}\boldsymbol{f}$ exist). The discontinuity is now encoded in the multiplier λ.

- E.g., convex combination (see Filippov [29, Def. 4a, pp. 50–52])

$$\dot{\boldsymbol{x}} = \tfrac{1}{2}(1+\lambda)\boldsymbol{f}^+(\boldsymbol{x}) + \tfrac{1}{2}(1-\lambda)\boldsymbol{f}^-(\boldsymbol{x}); \tag{1.11}$$

- E.g., nonlinear combination (see Jeffrey [59])

$$\dot{\boldsymbol{x}} = \tfrac{1}{2}(1+\lambda)\boldsymbol{f}^+(\boldsymbol{x}) + \tfrac{1}{2}(1-\lambda)\boldsymbol{f}^-(\boldsymbol{x}) + (\lambda^2-1)\boldsymbol{g}(\boldsymbol{x},\lambda). \tag{1.12}$$

In both examples $\boldsymbol{f}(\boldsymbol{x};\pm 1) \equiv \boldsymbol{f}^\pm(\boldsymbol{x})$. The function $\boldsymbol{g}$ can be any finite-valued vector field. The multiplier λ in either case satisfies (1.10).

The term $(\lambda^2-1)\boldsymbol{g}(\boldsymbol{x},\lambda)$ is called **hidden**, because it vanishes for $\boldsymbol{x} \notin \Sigma$ (when $\lambda = \pm 1$). (In fact it is so hidden that you won't find it in most other courses or texts on piecewise-smooth dynamics to date, and until 2013, Filippov's convex combination alone was considered standard). We'll generalize all of this for more complex switching later.

The geometry of the convex combination and the hidden term $(\lambda^2-1)\boldsymbol{g}$ are illustrated in fig. 1.13. The convex combination traces out a linear path in space along which the endpoint of the vector $\boldsymbol{f}$ might lie for a point on the discontinuity. The hidden term generalizes this by displacing the line to follow a curve.

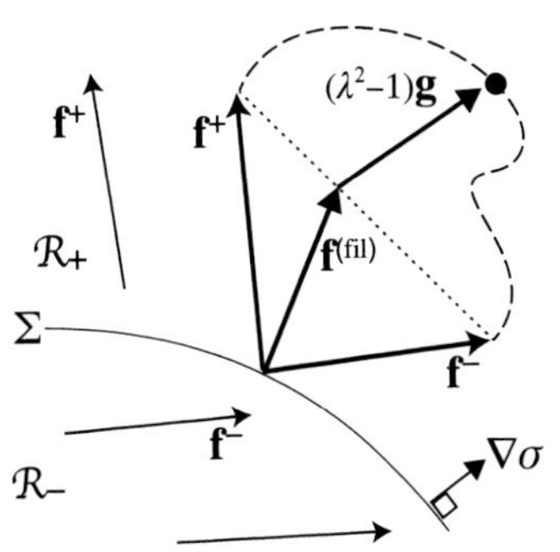

Figure 1.13: The Filippov convex combination (dotted line) written $\boldsymbol{f}^{(\mathrm{fil})} = \frac{1}{2}(1+\lambda)\boldsymbol{f}^{+} + \frac{1}{2}(1-\lambda)\boldsymbol{f}^{-}$, and a non-convex combination (dashed curve) formed by hidden term $(\lambda^2-1)\boldsymbol{g}$.

Example 1.3.4. Consider the vector field

$$\boldsymbol{f}(\boldsymbol{x};\lambda) = \tfrac{1}{2}(1+\lambda)(2,1) + \tfrac{1}{2}(1-\lambda)(-1,2) + (\lambda^2-1)(0,-1) \tag{1.13}$$

with a switching multiplier $\lambda = \operatorname{sign}(x_1 + x_2)$. This switches between $(2,1)$ and $(-1,2)$ across $\sigma = x_1 + x_2$. The convex and nonlinear combinations are shown in fig. 1.14.

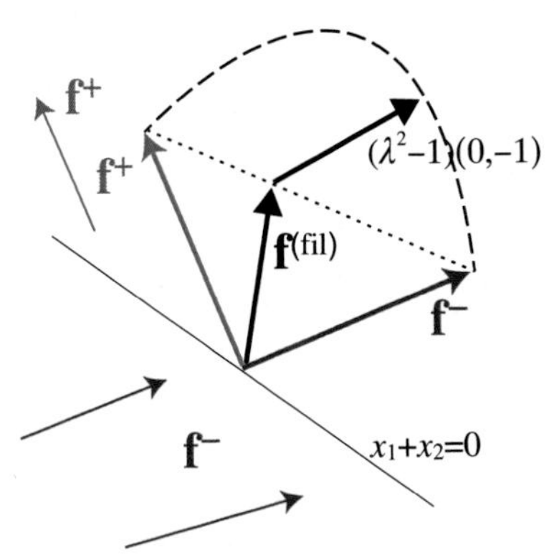

Figure 1.14: The Filippov convex combination (dotted line) and a non-convex combination (dashed curve) for theorem 1.3.4 with quadratic dependence on λ.

1.4 Types of dynamics

Filippov's theory tells us that a family of solutions of the equations, the flow, exists. What do the solutions look like?

1.4.1 The inclusion ... solutions

We concatenate smooth segments of solutions of $\dot{\boldsymbol{x}} = \mathcal{F}$ in the regions $\mathcal{R}_i$ and on Σ (shown in fig. 1.10), to form continuous curves that preserve the direction of time (illustrated in fig. 1.15).

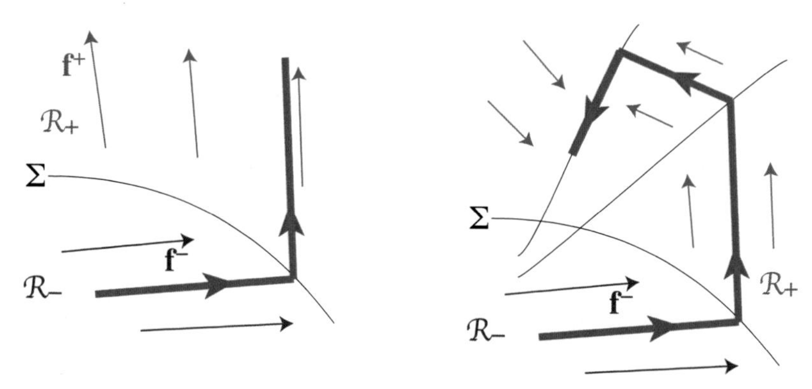

Figure 1.15: A crossing solution (left) corresponding to the inclusion from fig. 1.10 and fig. 1.13. We can concatenate over many discontinuities (right).

Definition 1.4.1. An **orbit** of the piecewise-smooth system through a point $\boldsymbol{x}_0$ is a maximal ('longest possible') concatenation of trajectories through $\boldsymbol{x}_0$.

The phrase 'longest possible' here is usually as interpreted as a continuous trajectory $\boldsymbol{x}(t)$ existing over a time interval $t \in (t_1, t_2)$ which is the largest possible interval, e.g., typically $t \in \mathbb{R}$ for $\boldsymbol{x} \in \mathbb{R}^n$, but possibly a finite interval if only a finite space is considered, e.g., $\boldsymbol{x} \in \mathcal{U} \subset \mathbb{R}^n$ where $\boldsymbol{x}(t_1)$ and $\boldsymbol{x}(t_2)$ lie on the boundary of $\mathcal{U}$ and $\boldsymbol{x}(t) \in \mathcal{U}$ for $t \in (t_1, t_2)$.

We can solve the equations $\dot{\boldsymbol{x}} = \boldsymbol{f}^i(\boldsymbol{x})$ for the flow inside the regions $\mathcal{R}_i$. We can now use the combination to find the flow on Σ.

A family of solutions parameterized by initial conditions $\boldsymbol{x}_0$ forms a flow $\Phi_t(\boldsymbol{x}_0)$ defined by

$$\boldsymbol{x}(t) = \Phi_t(\boldsymbol{x}_0): \quad \frac{d}{dt}\Phi_t(\boldsymbol{x}_0) = \boldsymbol{f}\left(\Phi_t(\boldsymbol{x}_0); \lambda\right), \quad \Phi_0(\boldsymbol{x}_0) = \boldsymbol{x}_0.$$

We may write the constituent flows in the regions $\mathcal{R}_+$, $\mathcal{R}_-$, as Φ_t^+, Φ_t^-, where

$$\frac{d}{dt}\Phi_t^{\pm}(\boldsymbol{x}_0) = \boldsymbol{f}\left(\Phi_t(\boldsymbol{x}_0); \pm 1\right).$$

1.4.2 Crossing and sliding

Let Φ_t^+, Φ_t^-, Φ_t^{Σ}, be the flows on $\mathcal{R}_+$, $\mathcal{R}_-$, Σ, respectively. At Σ:

- an orbit can **cross** through Σ at time τ.

  To cross from $\mathcal{R}_-$ to $\mathcal{R}_+$ at a point $\boldsymbol{x}_1 = \Phi^-_\tau(\boldsymbol{x}_0) \in \Sigma$ where $\sigma(\boldsymbol{x}_1) = 0$, it is *necessary* that $\dot\sigma(\Phi^-_{\tau-\delta t}(\boldsymbol{x}_0))$ and $\dot\sigma(\Phi^+_{\tau+\delta t}(\boldsymbol{x}_0))$ have the same sign, where δt is a small increment of time. Similarly to cross from $\mathcal{R}_+$ to $\mathcal{R}_-$ at a point $\boldsymbol{x}_1 = \Phi^+_\tau(\boldsymbol{x}_0) \in \Sigma$ where $\sigma(\boldsymbol{x}_1) = 0$, it is *necessary* that $\dot\sigma(\Phi^+_{\tau-\delta t}(\boldsymbol{x}_0))$ and $\dot\sigma(\Phi^-_{\tau+\delta t}(\boldsymbol{x}_0))$ have the same sign. In either case, therefore, crossing requires

$$(\boldsymbol{f}^+ \cdot \nabla\sigma)(\boldsymbol{f}^- \cdot \nabla\sigma) > 0 \quad \text{on } \Sigma. \tag{1.14}$$

- an orbit can **slide** along Σ, where the flow Φ^Σ_t must satisfy

$$\boldsymbol{x}(t) = \Phi^\Sigma_t(\boldsymbol{x}_0): \quad \frac{d}{dt}\sigma(\boldsymbol{x}(t)) = 0 \quad \text{on } \Sigma \tag{1.15}$$

  In the convex combination (1.11), since $\dot\sigma = \dot{\boldsymbol{x}} \cdot \nabla\sigma = \boldsymbol{f} \cdot \nabla\sigma$, sliding requires that both vector fields $\boldsymbol{f}^\pm$ point towards (or both away from) Σ, therefore sliding occurs where

$$(\boldsymbol{f}^+ \cdot \nabla\sigma)(\boldsymbol{f}^- \cdot \nabla\sigma) < 0 \quad \text{on } \Sigma, \tag{1.16}$$

and satisfies

$$0 = \dot\sigma = \dot{\boldsymbol{x}} \cdot \nabla\sigma = \left\{\frac{1+\lambda}{2}\boldsymbol{f}^+(\boldsymbol{x}) + \frac{1-\lambda}{2}\boldsymbol{f}^-(\boldsymbol{x})\right\} \cdot \nabla\sigma \tag{1.17}$$

$$\Rightarrow \quad \lambda = \lambda^\Sigma \equiv \frac{(\boldsymbol{f}^- + \boldsymbol{f}^+)\cdot\nabla\sigma}{(\boldsymbol{f}^- - \boldsymbol{f}^+)\cdot\nabla\sigma} \tag{1.18}$$

$$\Rightarrow \quad \dot{\boldsymbol{x}} = \boldsymbol{f}^\Sigma \equiv \frac{(\boldsymbol{f}^- \cdot \nabla\sigma)\boldsymbol{f}^+ - (\boldsymbol{f}^+ \cdot \nabla\sigma)\boldsymbol{f}^-}{(\boldsymbol{f}^- - \boldsymbol{f}^+)\cdot\nabla\sigma}. \tag{1.19}$$

In three dimensions this is compactly written as

$$\boldsymbol{f}^\Sigma \equiv \frac{\nabla\sigma \times (\boldsymbol{f}^+ \times \boldsymbol{f}^-)}{(\boldsymbol{f}^- - \boldsymbol{f}^+)\cdot\nabla\sigma}. \tag{1.20}$$

For sliding to exist we must have $\lambda^\Sigma \in (-1,+1)$ by (1.10), and indeed it is easy to show that (1.18) satisfies this if and only if (1.16) is satisfied. Sliding can occur outside (1.16) for a nonlinear switching system like (1.12), occurring wherever $\dot\sigma = \sigma = 0$ can be satisfied for $[\epsilon]\,(-1,+1)$.

Later it will be useful to observe that, from (1.19), we have

$$\boldsymbol{f}^\Sigma = \boldsymbol{f}^+ \quad \text{if } \boldsymbol{f}^+ \cdot \nabla\sigma = 0, \tag{1.21}$$

$$\boldsymbol{f}^\Sigma = \boldsymbol{f}^- \quad \text{if } \boldsymbol{f}^- \cdot \nabla\sigma = 0, \tag{1.22}$$

meaning that when the sliding vector field becomes equal to $\boldsymbol{f}^+$ or $\boldsymbol{f}^-$ at points where their normal components vanish, i.e., where they lie tangent to Σ. These two scenarios correspond to $\lambda^\Sigma = +1$ and $\lambda^\Sigma = -1$ by (1.18), and therefore also apply to (1.12).

Defining sliding dynamics therefore generally requires solving for the value of λ that provides motion along $\sigma = 0$:

Definition 1.4.2. The sliding dynamics is given by

$$\dot{\boldsymbol{x}} = \boldsymbol{f}^{\Sigma}(\boldsymbol{x}) \quad \text{on } \sigma(\boldsymbol{x}) = 0 \tag{1.23}$$

where $\boldsymbol{f}^{\Sigma}(\boldsymbol{x})$ is the **sliding vector field**, given by

$$\boldsymbol{f}^{\Sigma}(\boldsymbol{x}) \equiv \boldsymbol{f}(\boldsymbol{x};\lambda^{\Sigma}) \quad \text{such that } \boldsymbol{f}(\boldsymbol{x};\lambda^{\Sigma})\cdot\nabla\sigma = 0 \text{ and } \lambda^{\Sigma} \in (-1,+1) \tag{1.24}$$

on $\sigma(\boldsymbol{x}) = 0$.

Example 1.4.3. (i) The orbit

$$\boldsymbol{x}(t) = \begin{cases} \Phi_t^-(\boldsymbol{x}_0) & \text{if } t < \tau, \\ \Phi_{t-\tau}^+(\boldsymbol{x}_1) & \text{if } t > \tau, \end{cases} \tag{1.25}$$

crosses from $\mathcal{R}_-$ to $\mathcal{R}_+$ at time τ;

(ii) The orbit

$$\boldsymbol{x}(t) = \begin{cases} \Phi_t^-(\boldsymbol{x}_0) & \text{if } t < \tau, \\ \Phi_{t-\tau}^{\Sigma}(\boldsymbol{x}_1) & \text{if } t > \tau, \end{cases} \tag{1.26}$$

starts in $\mathcal{R}_-$, then **sticks** to Σ at time τ, sliding thereafter; where $\Phi_t^{\pm}$ are the flows of $\dot{\boldsymbol{x}} = \boldsymbol{f}^{\pm}$, and $\boldsymbol{x}_1 = \phi_\tau^-(\boldsymbol{x}_0) \in \Sigma$.

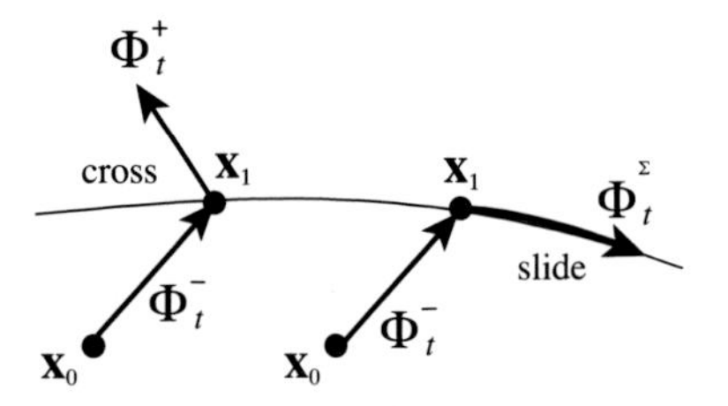

Figure 1.16: Examples of crossing and sliding.

Example 1.4.4 (Crossing)**.** Consider the system

$$(\dot{x}_1, \dot{x}_2) = (2+\lambda, 1), \qquad \lambda = \text{sign}(x_1), \tag{1.27}$$

so $\sigma(\boldsymbol{x}) = x_1$. Let $\tau > 0$. See fig. 1.17.

There is an obvious crossing solution

$$x_1(t) = (t-\tau)(2+\text{sign}(t-\tau)), \quad x_2(t) = t + x_{20}. \tag{1.28}$$

Does sliding ($\dot{\sigma} = \sigma = 0$) occur on Σ ($x_1 = 0$)? To find out solve

$$\dot{\sigma} = \dot{x}_1 = 0 \quad \Rightarrow \quad \lambda = -2 \notin (-1,+1) \quad \Rightarrow \quad \text{no sliding.} \tag{1.29}$$

The value of λ that would give sliding along Σ lies outside of the allowed range $(-1,+1)$, so no sliding modes exist.

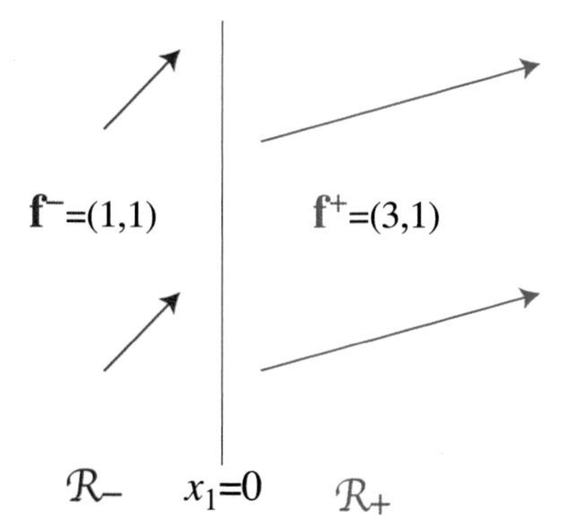

Figure 1.17: A piecewise constant example of crossing. We will revisit this example later to see it is not as trivial as it first appears.

Example 1.4.5 (Sliding). Consider the system

$$(\dot{x}_1, \dot{x}_2) = (-\lambda, 1), \qquad \lambda = \text{sign}(x_1), \tag{1.30}$$

so $\sigma(\boldsymbol{x}) = x_1$. Let $\tau > 0$; see fig. 1.18. Crossing is impossible, as both vector fields points towards Σ ($x_1 = 0$). So, surely the only possible motion is sliding? Solve

$$\dot{\sigma} = \dot{x}_1 = 0 \quad \Rightarrow \quad \lambda^{\Sigma} = 0 \tag{1.31}$$

which lies inside $(-1, +1)$, and substituting back into the original system

$$\Rightarrow \quad (\dot{x}_1, \dot{x}_2) = (0, 1). \tag{1.32}$$

An example of a sticking orbit is

$$x_1(t) = (\tau - t\,\text{sign}(\tau))\,\text{step}(\tau - t), \quad x_2(t) = t + x_{20}. \tag{1.33}$$

1.4.3 The local singularities

Any of the constituent systems $\dot{\boldsymbol{x}} = \boldsymbol{f}^{\pm}(\boldsymbol{x})$ may have an equilibrium where $\boldsymbol{f}^{\pm}(\boldsymbol{x}) = 0$, or in combination notation, where $\dot{\boldsymbol{x}} = \boldsymbol{f}(\boldsymbol{x}; \pm 1) = 0$.

Two new singularities arise at a (simple) switching surface:

(i) Sliding ("pseudo") equilibria are points where $\boldsymbol{f}^{\pm}(\boldsymbol{x}) \neq 0$ but $\boldsymbol{f}^{\Sigma}(\boldsymbol{x}) = 0$, i.e.,

$$\boldsymbol{f}(\boldsymbol{x}; \lambda^{\Sigma}) = 0 \;\text{ with }\; \sigma(\boldsymbol{x}) = 0. \tag{1.34}$$

In the convex combination these happen where $\boldsymbol{f}^{\pm}$ are in opposition, since for some $\mu > 0$

$$\begin{aligned} \boldsymbol{f}^{+} = -\mu \boldsymbol{f}^{-} \quad \Rightarrow \quad \dot{\boldsymbol{x}} = \boldsymbol{f}^{\Sigma} &= \frac{(\boldsymbol{f}^{-} \cdot \nabla\sigma)\boldsymbol{f}^{+} - (\boldsymbol{f}^{+} \cdot \nabla\sigma)\boldsymbol{f}^{-}}{(\boldsymbol{f}^{-} - \boldsymbol{f}^{+}) \cdot \nabla\sigma} \\ &= \frac{(\boldsymbol{f}^{-} \cdot \nabla\sigma)(-\mu \boldsymbol{f}^{-}) - (-\mu \boldsymbol{f}^{-} \cdot \nabla\sigma)\boldsymbol{f}^{-}}{(\boldsymbol{f}^{-} + \mu \boldsymbol{f}^{-}) \cdot \nabla\sigma} = 0. \end{aligned} \tag{1.35}$$

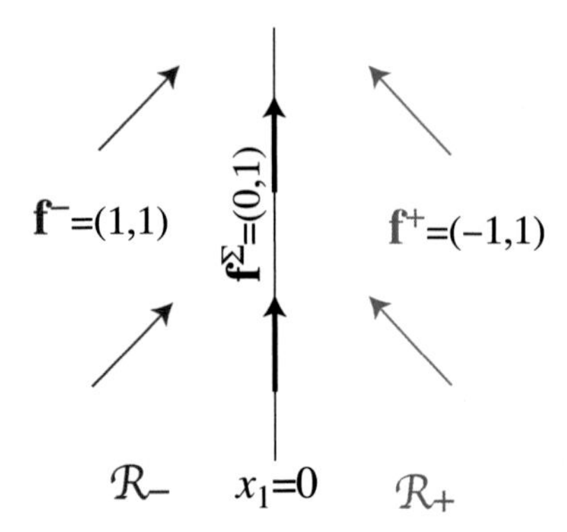

Figure 1.18: A piecewise constant example of sliding.

These are stationary points of the sliding flow on the switching surface.

(ii) Tangencies of the vector field to Σ are points where:

$$\boldsymbol{f}^+ \cdot \nabla\sigma = 0 \quad \Rightarrow \quad \dot{\boldsymbol{x}} = \boldsymbol{f}^+ \tag{1.36}$$

or similarly for $\boldsymbol{f}^-$. These are stationary points with respect to the switching surface, of the flows outside the surface. It is easy seen using (1.11) or (1.12) that they correspond to points where $\boldsymbol{f}(\boldsymbol{x};\pm1) = 0$, and so for example $\lambda^\Sigma = \pm1$ in (1.18).

Example 1.4.6 (Sliding equilibria). In two dimensions consider $\lambda = \operatorname{sign} x_1$ with

$$(\dot{x}_1, \dot{x}_2) = \tfrac{1}{2}(1+\lambda)(-1,-x_2) + \tfrac{1}{2}(1-\lambda)(1,b) \tag{1.37}$$

or in three dimensions $\lambda = \operatorname{sign} x_1$ with

$$(\dot{x}_1, \dot{x}_2, \dot{x}_3) = \tfrac{1}{2}(1+\lambda)(-1, x_3 - x_2, -x_2) + \tfrac{1}{2}(1-\lambda)(1,b,c). \tag{1.38}$$

Clearly the $\lambda = +1$ system in $x_1 > 0$ and the $\lambda = -1$ system in $x_1 < 0$ are non-vanishing. In both cases, to find sliding modes we solve $0 = \dot{\sigma} = \dot{x}_1 = \frac{1}{2}(1+\lambda)(-1) + \frac{1}{2}(1-\lambda)(+1) = \lambda$. Hence in (1.37)

$$\lambda^\Sigma = 0 \quad \Rightarrow \quad \dot{x}_2 = \tfrac{1}{2}(b - x_2),$$

with a sliding equilibrium at $x_2 = b$; and in (1.38)

$$\lambda^\Sigma = 0 \quad \Rightarrow \quad (\dot{x}_2, \dot{x}_3) = \tfrac{1}{2}(b - x_2 + x_3, c - x_2),$$

with a sliding equilibrium at $(x_2, x_3) = (c, c-b)$; see fig. 1.19.

Note in the two systems that the directions of the upper and lower vector fields (above and below $x_1 = 0$), at the equilibria are given by $(-1,-b)$ and $(1,b)$ in (1.37) and $(-1,-b,-c)$ and $(1,b,c)$ in (1.38), i.e., they are antiparallel there (their magnitudes are not significant).

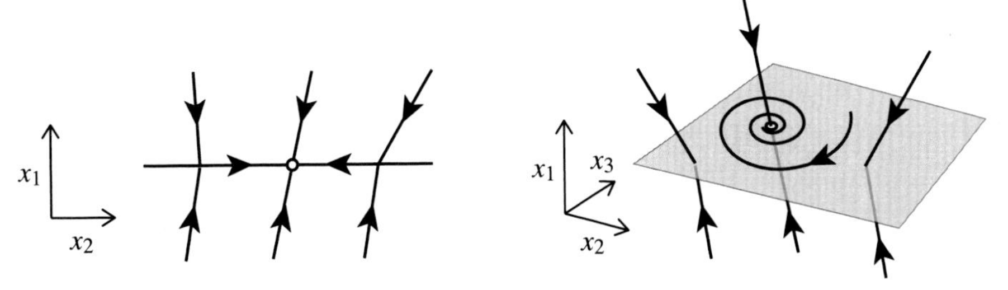

Figure 1.19: Sliding equilibria in two or three dimensions. For (1.37) to (1.38) the x_1 direction is vertical.

Example 1.4.7 (Tangencies). Figure 1.20 below shows quadratic tangencies of types we call 'visible' (curving away from Σ) or 'invisible' (curving towards Σ).

In higher dimensions, sets of visible and invisible tangencies meet at higher order tangencies, like the cubic tangency, called a 'cusp', illlustrated in fig. 1.21.

The examples shown in fig. 1.20 and fig. 1.21 are from the equations

$$\begin{aligned}
(\dot{x}_1, \dot{x}_2) &= \tfrac{1}{2}(1+\lambda)(-1, -x_1) + \tfrac{1}{2}(1-\lambda)(0,1), && \lambda = \mathrm{sign}(x_2), \\
(\dot{x}_1, \dot{x}_2) &= \tfrac{1}{2}(1+\lambda)(+1, -x_1) + \tfrac{1}{2}(1-\lambda)(0,1), && \lambda = \mathrm{sign}(x_2), \\
(\dot{x}_1, \dot{x}_2, \dot{x}_3) &= \tfrac{1}{2}(1+\lambda)(1, 0, x_1^2 + x_2) + \tfrac{1}{2}(1-\lambda)(0,0,1), && \lambda = \mathrm{sign}(x_3),
\end{aligned}$$

but we could also have cubic, quartic, etc. order.

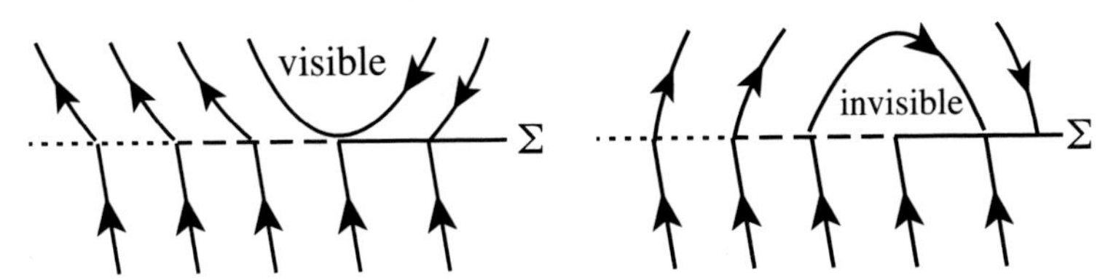

Figure 1.20: Visible and invisible tangencies in the upper vector field. On the switching surface Σ, the full line indicates the sliding region found using a convex combination, but nonlinear dependence on λ can let this can bleed out as indicated by the dashed region, while the remainder is crossing (dotted).

At the simple ('fold') tangencies in fig. 1.20 we have that the upper ($\lambda = +1$ vector field) satisfies $\dot{x}_2 = -x_1$ which vanishes at x_1, creating a tangency. The sign of the second derivative determines its curvature, in the first example $\ddot{x}_2 = -\dot{x}_1 = +1$ implies the flow is curving upwards, creating a *visible* tangency, and in the second example $\ddot{x}_2 = -\dot{x}_1 = -1$ implies the flow is curving downwards, creating an *invisible* tangency

At the cusp, for the upper vector field we have $\dot{x}_3 = x_1^2 + x_2$ which vanishes on $x_3 = 0$ along a curve $x_1 = \pm\sqrt{-x_2}$ (existing for $x_2 < 0$). The second derivative is $\ddot{x}_3 = 2\dot{x}_1 x_1 + \dot{x}_2 = 2x_1 = \pm 2\sqrt{-x_2}$, meaning the half-curve $x_1 = +\sqrt{-x_2}$ is a family of visible folds while the half-curve $x_1 = -\sqrt{-x_2}$ is a family of invisible folds. The point $x_1 = x_2 = x_3 = 0$ where $\ddot{x}_3 = 0$ is the cusp.

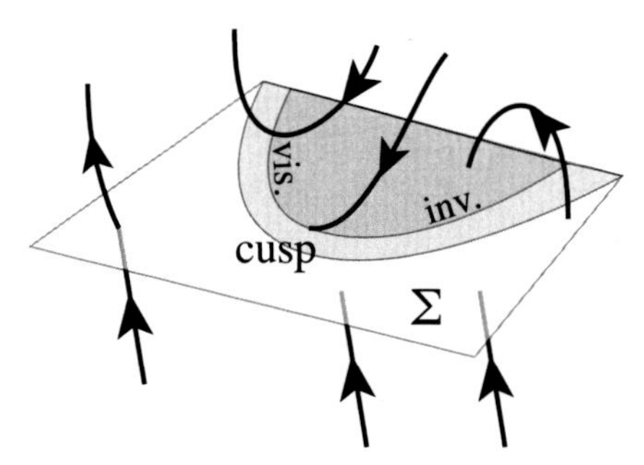

Figure 1.21: A cusp in the upper vector field. On the switching surface Σ, the dark shaded region indicates the sliding region found using a convex combination, but nonlinear dependence on λ can let this can bleed out as indicated by the light shaded region, while the remainder is crossing (unshaded).

Above we gave a necessary condition for crossing, but it does not guarantee crossing (it is not necessary *and sufficient*). The region indicating by the full line in fig. 1.20 and by shading fig. 1.21 on Σ shows where both vector fields point towards the surface, $\boldsymbol{f}^+ \cdot \nabla\sigma < 0 < \boldsymbol{f}^- \cdot \nabla\sigma$, so sliding *must* occur. In the remaining region trajectories may cross through (and in Filippov's convention they *do* cross through), but in nonlinear combinations it is possible for the sliding region to 'bleed out' into the crossing region, indicated by the dashed/light-shaded region on Σ.

We've hinted above that crossing or sliding on a switching surface depends on convexity, i.e., whether the dependence on λ is linear. Most courses would assume linearity and hasten onward. Let's hold back a little and consider nonlinear dependence a little further.

1.4.4 Crossing and sliding – examples II

Let's revisit the crossing example in (1.27) above and add a 'hidden' term.

Example 1.4.8 (Nonlinear λ dependence and crossing)**.** Consider

$$(\dot{x}_1, \dot{x}_2) = (2 + \lambda, 1) + 2(\lambda^2 - 1, 0), \quad \lambda = \text{sign}(x_1), \tag{1.39}$$

so $\sigma(\boldsymbol{x}) = x_1$. Let $\tau > 0$. The obvious crossing solution is again $x_1(t) = (t - \tau)(2 + \text{sign}(t - \tau))$, $x_2(t) = t + x_{20}$, but turns out to be wrong in the presence of the nonlinear term. Let's look for sliding modes by solving $\dot{\sigma} = \sigma = 0$ on Σ ($x_1 = 0$):

$$0 = \dot{\sigma} = \dot{x}_1 = \lambda + 2\lambda^2 \quad \Rightarrow \quad \lambda^\Sigma = -\tfrac{1}{2} \text{ or } 0. \tag{1.40}$$

Thus, there are two sliding modes. It turns out that either modes gives sliding dynamics $(\dot{x}_1, \dot{x}_2) = (0, 1)$ for this simple example.

Now we seem to have three possible solutions at Σ: the orbit may cross, or it may follow one of two sliding modes. This kind of ambiguity leads to paradoxes in physics if incorrectly handed. So which is right?

To find out we have to look closer at how λ varies on the interval -1 to $+1$.

1.5 Switching layers

1.5.1 The dynamics of λ

Switching layers resolve the jump in λ into a continuous (but infinitesimal time) dynamics that takes place in an infintesimally short time, with invariant sets corresponding to the sliding dynamics $\dot{\sigma} = 0$. We define

$$\varepsilon\dot{\lambda} = \boldsymbol{f} \cdot \nabla\sigma \quad \text{for } \varepsilon \to 0. \tag{1.41}$$

The argument for (1.41) begins with noting that the multiplier λ is a function only of σ, so it makes sense that $\dot{\lambda}$ be determined by $\dot{\sigma}$. Let $\lambda = \lambda(\sigma/\varepsilon)$ for some $\varepsilon \geq 0$, then noting $\dot{\sigma} = \boldsymbol{f} \cdot \nabla$,

$$\dot{\lambda} = \lambda'\dot{\sigma}/\varepsilon = \lambda'\boldsymbol{f} \cdot \nabla\sigma/\varepsilon \quad \Rightarrow \quad \tilde{\varepsilon}\dot{\lambda} = \boldsymbol{f} \cdot \nabla\sigma \tag{1.42}$$

defining $\tilde{\varepsilon} = \varepsilon/\lambda'$. We assume that λ' is sufficiently well behaved (non-vanishing on an open interval $|\sigma| < \varepsilon$ such that $\tilde{\varepsilon} \to 0$ as $\varepsilon \to 0$, see [62]). Since only $\varepsilon \to 0$ is of interest, we then drop the tilde in (1.42).

Letting $\varepsilon \to 0$ gives an instantaneous switch (instantaneous because the rate of change $\dot{\lambda} \sim 1/\varepsilon$ is then infinitely large), so in piecewise-smooth systems, this is the limit we're interested in.

The equilibrium of (1.41) is just the sliding mode, since

$$\dot{\lambda} = 0 \quad \Leftrightarrow \quad \dot{\sigma} = 0. \tag{1.43}$$

Let us apply this now to the previous problem.

Example 1.5.1 (Nonlinear λ ... continued). On $x_1 = 0$ let

$$\varepsilon\dot{\lambda} = 2 + \lambda + 2(\lambda^2 - 1). \tag{1.44}$$

This has two equilibria, at $\lambda = \lambda^\Sigma = -\frac{1}{2}$ or 0. Since

$$\left.\frac{\partial\dot{\lambda}}{\partial\lambda}\right|_{\lambda=\lambda^\Sigma} = 1 + 4\lambda^\Sigma = \begin{cases} -1 & \text{if } \lambda^\Sigma = -\frac{1}{2}, \\ +1 & \text{if } \lambda = 0, \end{cases} \tag{1.45}$$

the solution $\lambda^\Sigma = -\frac{1}{2}$ is attractive (the other is repelling), so this is the solution the flow follows.

1.5.2 The switching layer

When we take the dynamical equation on λ above, what we actually do is magnify the *surface* $\sigma = 0$ into a *layer* over which $\lambda \in (-1, +1)$ on $\sigma = 0$.

Take coordinates so that $\sigma = x_1$, then:

Definition 1.5.2. The **switching layer** on $x_1 = 0$ is

$$(\lambda, x_2, \ldots, x_n) \in (-1, +1) \times \mathbb{R}^{n-1}. \tag{1.46}$$

Figure 1.22 shows the switching layer for (1.27), in which crossing occurs.

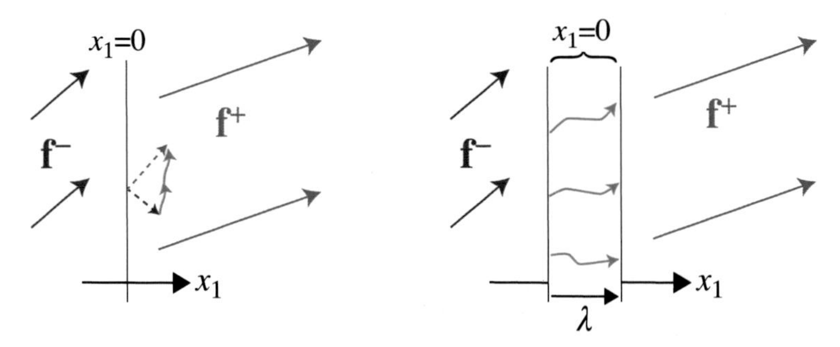

Figure 1.22: A simple crossing region, showing the jump in the vector field (left), blown up to reveal the switching layer (right).

Figure 1.23 shows the corresponding picture for (1.44), in which crossing is prohibited by hidden terms, determined by nonlinear dependence on the switching multiplier λ.

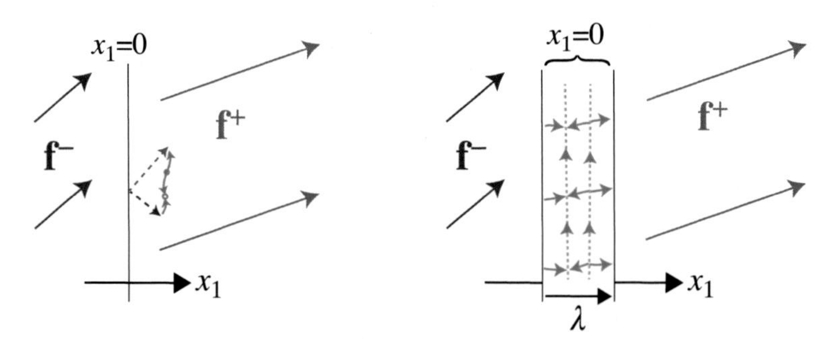

Figure 1.23: A nonlinear sliding region, showing the jump in the vector field (left), blown up to reveal the switching layer (right) with attracting and repelling sliding modes.

For rigorous theory concerning switching layers see [62, 79]. The theory is at an early stage of development, but has already begun helping resolve some of the outstanding problems in piecewise-smooth dynamics, and we'll see some examples later.

Inside the switching layer the variation is given by the two time-scale system obtained by putting $\dot{\boldsymbol{x}} = \boldsymbol{f}$ together with $\varepsilon\lambda = \boldsymbol{f}\cdot\nabla\sigma$ on $\sigma = 0$. In coordinates where $\sigma = x_1$:

Definition 1.5.3. The switching layer system is

$$\begin{aligned} \varepsilon\dot{\lambda} &= f_1(0, x_2, \ldots, x_n; \lambda), \quad && \lambda \in (-1, +1), \\ \dot{x}_i &= f_i(0, x_2, \ldots, x_n; \lambda), \quad && i = 2, \ldots, n, \end{aligned}$$

in terms of an infinitesimal $\varepsilon \geq 0$ in the limit $\varepsilon \to 0$.

1.5.3 Layer variables

The layer system allows us to stretch space to peer inside the discontinuity. But it also gives something more, a way to extend certain local methods from smooth dynamical theory, like linearization, to discontinuities. To do this we will need a re-scaled variable, called the layer variable.

Definition 1.5.4. In a system where $\lambda = \text{sign}(x_1)$, the switching **layer variable** is the vector

$$\boldsymbol{\xi} = (\xi_1, \xi_2, \ldots, \xi_n) = (\varepsilon\lambda, x_2, \ldots, x_n) \tag{1.47}$$

(i.e., the vector $\boldsymbol{\xi}$ given by replacing x_1 by $\varepsilon\lambda$ in coordinates where $\sigma = x_1$).

1.5.4 Layer variables – examples

Example 1.5.5 (I. Linearization). The planar system

$$\begin{pmatrix} \dot{x}_1 \\ \dot{x}_2 \end{pmatrix} = \begin{pmatrix} -\lambda \\ c - x_2 - (c + x_2)\lambda \end{pmatrix} = \begin{cases} \begin{pmatrix} -1 \\ -2x_2 \end{pmatrix} & \text{if } x_1 > 0, \\ \begin{pmatrix} 1 \\ 2c \end{pmatrix} & \text{if } x_1 < 0, \end{cases}$$

has no equilibria for $x_1 \neq 0$. The switching layer system on $x_1 = 0$ is

$$\begin{pmatrix} \varepsilon\dot{\lambda} \\ \dot{x}_2 \end{pmatrix} = \begin{pmatrix} -\lambda \\ c - x_2 - (c + x_2)\lambda \end{pmatrix}. \tag{1.48}$$

This has a sliding equilibrium at $(\lambda, x_2) = (0, c)$. Is it an attractor? What are it's eigenvalues and eigenvectors?

In layer variables on $x_1 = 0$ we have

$$\begin{pmatrix} \dot{\xi}_1 \\ \dot{\xi}_2 \end{pmatrix} = \begin{pmatrix} -\xi_1/\varepsilon \\ c - \xi_2 - (c + \xi_2)\xi_1/\varepsilon \end{pmatrix}. \tag{1.49}$$

Let us find the Jacobian in the layer variables $\boldsymbol{\xi} = (\xi_1, \xi_2) = (\varepsilon\lambda, x_2)$, evaluated at the equilibrium:

$$\underline{\underline{J}} = \begin{pmatrix} \frac{\partial \dot{\xi}_1}{\partial \xi_1} & \frac{\partial \dot{\xi}_1}{\partial \xi_2} \\ \frac{\partial \dot{\xi}_2}{\partial \xi_1} & \frac{\partial \dot{\xi}_2}{\partial \xi_2} \end{pmatrix} = -\frac{1}{\varepsilon}\begin{pmatrix} 1 & 0 \\ c+\xi_2 & \xi_1+\varepsilon \end{pmatrix} = -\frac{1}{\varepsilon}\begin{pmatrix} 1 & 0 \\ 2c & \varepsilon \end{pmatrix} \tag{1.50}$$

with eigenvalues ν_i and eigenvectors $\boldsymbol{v}_i$ (solutions of $\underline{\underline{J}}\boldsymbol{v}_i = \nu_i \boldsymbol{v}_i$) as $\varepsilon \to 0$:

$$\begin{aligned} \nu_1 &= -1/\varepsilon \to -\infty : & \boldsymbol{v}_1 &\to (1, 2c)^{\mathsf{T}}, \\ \nu_2 &= -1 : & \boldsymbol{v}_2 &= (0,\ 1\ )^{\mathsf{T}} \quad \Leftrightarrow \text{ in line of } \Sigma. \end{aligned} \tag{1.51}$$

Both are attracting (agreeing with this being a node, since $\det \underline{\underline{J}} = 1/\varepsilon > 0$). Attraction along the $(0,1)^{\mathsf{T}}$ direction is at unit rate $x_2 - c \sim (x_{20} - c)e^{-t}$. Attraction along the $(1,2c)^{\mathsf{T}}$ direction is infinitely fast, $\xi_1 \sim \xi_{10}e^{-t/\varepsilon}$, so while it takes a finite time to reach $x_1 = 0$, once there, in any time instant t we find ξ_1 contracts immediately to the equilibrium at $e^{-t/\varepsilon} \to 0$ as $\varepsilon \to 0$.

Example 1.5.6 (II. Linearization). The planar system

$$\begin{pmatrix} \dot{x}_1 \\ \dot{x}_2 \end{pmatrix} = \begin{pmatrix} 1 - x_2 - \lambda(1+x_2) \\ c - 1 - \lambda(c+1) \end{pmatrix} = \begin{cases} 2\begin{pmatrix} -x_2 \\ -1 \end{pmatrix} & \text{if } x_1 > 0, \\ 2\begin{pmatrix} 1 \\ c \end{pmatrix} & \text{if } x_1 < 0, \end{cases}$$

has no equilibria for $x_1 \neq 0$. The switching layer system is

$$\begin{pmatrix} \varepsilon\dot{\lambda} \\ \dot{x}_2 \end{pmatrix} = \begin{pmatrix} 1 - x_2 - \lambda(1+x_2) \\ c - 1 - \lambda(c+1) \end{pmatrix}. \tag{1.52}$$

This has an equilibrium at $(\lambda, x_2) = (\frac{c-1}{c+1}, \frac{1}{c})$. Is it an attractor? What are its eigenvalues/vectors?

Find the Jacobian in the layer variables $\boldsymbol{\xi} = (\xi_1, \xi_2) = (\varepsilon\lambda, x_2)$, evaluated at the equilibrium:

$$\underline{\underline{J}} = \begin{pmatrix} \frac{\partial \dot{\xi}_1}{\partial \xi_1} & \frac{\partial \dot{\xi}_1}{\partial \xi_2} \\ \frac{\partial \dot{\xi}_2}{\partial \xi_1} & \frac{\partial \dot{\xi}_2}{\partial \xi_2} \end{pmatrix} = -\frac{1}{\varepsilon}\begin{pmatrix} 1+\xi_2 & \xi_1+\varepsilon \\ c+1 & 0 \end{pmatrix} = -\frac{1}{\varepsilon}\begin{pmatrix} 1+\frac{1}{c} & \frac{2c\varepsilon}{c+1} \\ c+1 & 0 \end{pmatrix} \tag{1.53}$$

with eigenvalues and eigenvectors as $\varepsilon \to 0$:

$$\begin{aligned} \nu_1 &= \tfrac{c+1+\sqrt{R}}{-4c\varepsilon} \to -\infty : & \boldsymbol{v}_1 &\to (1, c)^{\mathsf{T}}, \\ \nu_2 &= \tfrac{c+1-\sqrt{R}}{-4c\varepsilon} \to \tfrac{c^2}{c+1} : & \boldsymbol{v}_2 &= (0,\ 1\ )^{\mathsf{T}} \quad \Leftrightarrow \text{ in line of } \Sigma, \end{aligned} \tag{1.54}$$

where $R = (c+1)^2 + 8c^3\varepsilon$.

Say $c > 0$. Then one of these eigendirections is attracting, and one repelling (implying a saddle, agreeing with $\det \underline{\underline{J}} = -2c/\varepsilon < 0$).

Repulsion along the $(0,1)^{\mathsf{T}}$ direction is as $x_2 - \frac{1}{c} \sim (x_{20} - \frac{1}{c})e^{+tc^2/(c+1)}$, and therefore takes infinite time to leave $x_2 = 1/c$.

Along the $(1,c)^{\mathsf{T}}$ direction the rate of attraction is infinitely fast, so the contraction to $\xi_1 = 0$ is instantaneous, and $\xi_1 - \frac{c-1}{c+1} \sim (\xi_{10} - \frac{c-1}{c+1})e^{-t(c+1)/2c\varepsilon}$ gives $\xi_1 \to \frac{c-1}{c+1}$ for $\varepsilon \to 0$.

We leave it for the reader to consider the case $c < 0$.

Example 1.5.7 (III. Linearization). The three-dimensional system

$$\begin{pmatrix} \dot{x}_1 \\ \dot{x}_2 \\ \dot{x}_3 \end{pmatrix} = \begin{cases} \begin{pmatrix} -1 \\ ax_2 + bx_3 \\ cx_2 + dx_3 \end{pmatrix} & \text{if } x_1 > 0, \\ \begin{pmatrix} 1 \\ e \\ 0 \end{pmatrix} & \text{if } x_1 < 0, \end{cases}$$

has no equilibria for $x_1 \neq 0$. Taking the convex combination, the switching layer system simplifies to

$$\begin{pmatrix} \varepsilon\dot{\lambda} \\ \dot{x}_2 \\ \dot{x}_3 \end{pmatrix} = \tfrac{1}{2}\begin{pmatrix} -2\lambda \\ (1+\lambda)(ax_2 + bx_3) + (1-\lambda)e \\ (1+\lambda)(cx_2 + dx_3) \end{pmatrix}. \tag{1.55}$$

This has an equilibrium at $(\lambda, x_2, x_3) = (0, -d, c)e/(ad - bc)$. In layer variables,

$$\begin{pmatrix} \dot{\xi}_1 \\ \dot{\xi}_2 \\ \dot{\xi}_3 \end{pmatrix} = \tfrac{1}{2\varepsilon}\begin{pmatrix} -2\xi_1 \\ (\varepsilon + \xi_1)(a\xi_2 + b\xi_3) + (\varepsilon - \xi_1)e \\ (\varepsilon + \xi_1)(c\xi_2 + d\xi_3) \end{pmatrix}, \tag{1.56}$$

whose Jacobian at the sliding equilibrium is

$$\underline{\underline{J}} = \tfrac{1}{2}\begin{pmatrix} -2/\varepsilon & 0 & 0 \\ -2e/\varepsilon & a & b \\ 0 & c & d \end{pmatrix} \tag{1.57}$$

with eigenvalues and corresponding eigenvectors as $\varepsilon \to 0$

$$\begin{array}{llll} \nu_1 = -1/\varepsilon \to -\infty : & \boldsymbol{v}_1 \to (1, e, 0)^{\mathsf{T}}, & & \\ \nu_2 = -\frac{a+d}{4} - \sqrt{R} : & \boldsymbol{v}_2 = (0, 2\nu_2 - d, c)^{\mathsf{T}} & \Leftrightarrow & \text{in plane of } \Sigma, \\ \nu_3 = -\frac{a+d}{4} + \sqrt{R} : & \boldsymbol{v}_3 = (0, 2\nu_3 - d, c)^{\mathsf{T}} & \Leftrightarrow & \text{in plane of } \Sigma. \end{array} \tag{1.58}$$

where $R = \left(\frac{a+d}{4}\right)^2 + bc - ad$.

The eigenvalues $\nu_{2,3}$ are finite and confined to the (x_2, x_3) plane of Σ, where the equilibrium is: a saddle if $ad - bc < 0$, a focus if $ad - bc > 0 > R$, a node if $ad - bc > 0$ & $R > 0$, attracting if $a + d < 0$ and repelling if $a + d > 0$.

The eigenvalue ν_1 tells us that the sliding region $x_1 = 0$ is attracting with an infinite rate (i.e., in zero time) along the direction $(1/e, 1, 0)$, i.e., along the directions of the two constituent vector fields $\boldsymbol{f}^{\pm}$ at the equilibrium.

We will now take a look at local analysis of a bifurcation of equilibria. First, let us recall the basic bifurcation of equilibria in a smooth system.

Example 1.5.8 (A saddle-node bifurcation in the plane)**.** Take a smooth system

$$\begin{pmatrix} \dot{x}_1 \\ \dot{x}_2 \end{pmatrix} = \begin{pmatrix} x_1^2 - c \\ -x_2 \end{pmatrix} \tag{1.59}$$

which has equilibria at $(x_1, x_2) = (\pm\sqrt{c}, 0)$, which exist only for $c > 0$.

The Jacobian is $\underline{\underline{J}} = \begin{pmatrix} \pm\sqrt{c} & 0 \\ 0 & -1 \end{pmatrix}$ at $(x_1, x_2) = (\pm\sqrt{c}, 0)$, with

$$\begin{array}{ll} \nu_1 = -1 : & \boldsymbol{v}_1 \to (0,1)^{\mathsf{T}}, \\ \nu_2 = \pm 2\sqrt{c} : & \boldsymbol{v}_2 = (1,0)^{\mathsf{T}}. \end{array} \tag{1.60}$$

The '+' gives a saddle at $(x_1, x_2) = (+\sqrt{c}, 0)$ and the '−' gives an attracting node at $(x_1, x_2) = (-\sqrt{c}, 0)$, fig. 1.24.

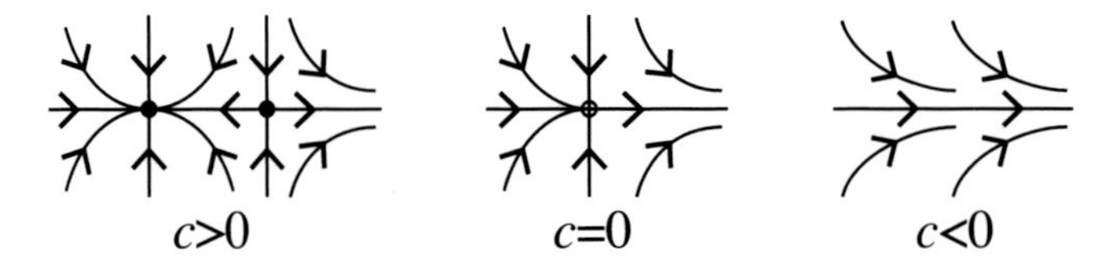

Figure 1.24: A saddle-node bifurcation in a smooth system.

A saddle-node bifurcation occurs as c changes sign: the two equilibria (for $c > 0$) annihilate each other (at $x = 0$) and leave a non-vanishing flow (for $c < 0$).

Let's contrast this with a similar boundary equilibrium bifurcation.

Example 1.5.9 (A discontinuity-induced saddle-node bifurcation)**.** Take a piecewise-smooth system

$$\begin{pmatrix} \dot{x}_1 \\ \dot{x}_2 \end{pmatrix} = \tfrac{1}{2}(1+\lambda)\begin{pmatrix} 1 \\ b \end{pmatrix} + \tfrac{1}{2}(1-\lambda)\begin{pmatrix} -\frac{1}{2}x_1 \\ -x_2 \end{pmatrix} \tag{1.61}$$

where $\lambda = \operatorname{sign}\sigma$ and $\sigma = x_1 - x_2 - c$, the bifurcation parameter is c, and b is a small nominal constant. For the layer system it helps to rotate coordinates, so let $(y_1, y_2) = (x_1 - x_2 - c, x_1 + x_2)/4$, giving

$$\begin{pmatrix} \dot{y}_1 \\ \dot{y}_2 \end{pmatrix} = \tfrac{1}{2}(1+\lambda)\begin{pmatrix} 1-b \\ 1+b \end{pmatrix} + \tfrac{1}{2}(1-\lambda)\begin{pmatrix} y_2 - 3y_1 - \frac{3}{4}c \\ y_1 - 3y_2 + \frac{1}{4}c \end{pmatrix} \tag{1.62}$$

in which the node lies at $(y_1, y_2) = (-\frac{c}{4}, 0)$, with Jacobian

$$\underline{\underline{J}} = \begin{pmatrix} -3 & 1 \\ 1 & -3 \end{pmatrix} \quad \Rightarrow \quad \begin{array}{ll} \nu_1 = -4: & \boldsymbol{v}_1 \to (-1,1)^\mathsf{T}, \\ \nu_2 = -2: & \boldsymbol{v}_2 = (1,1)^\mathsf{T}. \end{array} \tag{1.63}$$

The layer system on $y_1 = 0$ is

$$\begin{pmatrix} \varepsilon\dot{\lambda} \\ \dot{y}_2 \end{pmatrix} = \tfrac{1}{2}(1+\lambda)\begin{pmatrix} 1-b \\ 1+b \end{pmatrix} + \tfrac{1}{2}(1-\lambda)\begin{pmatrix} y_2 - 3y_1 - \frac{3}{4}c \\ y_1 - 3y_2 + \frac{1}{4}c \end{pmatrix} \tag{1.64}$$

with an equilibrium at $(\lambda, y_2) = \left(\frac{c+b-2}{c-b+2}, \frac{c(2+b)}{4(2-b)}\right)$, or in layer variables

$$\begin{pmatrix} \dot{\xi}_1 \\ \dot{\xi}_2 \end{pmatrix} = \tfrac{1}{2}(1+\xi_1/\varepsilon)\begin{pmatrix} 1-b \\ 1+b \end{pmatrix} + \tfrac{1}{2}(1-\xi_1/\varepsilon)\begin{pmatrix} \xi_2 - 3\xi_1 - \frac{3}{4}c \\ \xi_1 - 3\xi_2 + \frac{1}{4}c \end{pmatrix} \tag{1.65}$$

with Jacobian

$$\underline{\underline{J}} = \begin{pmatrix} \frac{\partial\dot{\xi}_1}{\partial\xi_1} & \frac{\partial\dot{\xi}_1}{\partial\xi_2} \\ \frac{\partial\dot{\xi}_2}{\partial\xi_1} & \frac{\partial\dot{\xi}_2}{\partial\xi_2} \end{pmatrix} = \begin{pmatrix} \frac{(b-1)(c-b+2)}{2\varepsilon(b-2)} & \frac{2-b}{c-b+2} \\ \frac{2+c+b+cb-b^2}{2\varepsilon(2-b)} & \frac{3(2-b)}{b-c-2} \end{pmatrix}, \tag{1.66}$$

whose eigenvalues and corresponding eigenvectors as $\varepsilon \to 0$ are

$$\begin{array}{llll} \nu_1 = \frac{1}{\varepsilon}\frac{(1-b)(c-b+2)}{2(2-b)} \to +\infty: & \boldsymbol{v}_1 \to (1-b, 1+b)^\mathsf{T}, & & \\ \nu_2 = \frac{2(2-b)^2}{(b-1)(c-b+2)} < 0: & \boldsymbol{v}_2 = (0,1)^\mathsf{T} & \Leftrightarrow & \text{in line of } \Sigma \end{array} \tag{1.67}$$

(for small c and b).

Hence this equilibrium is a saddle, with an infinite rate of repulsion along the $(1-b, 1+b)$ direction out of the switching surface, and asymptotic attraction along the vertical inside the surface, fig. 1.25.

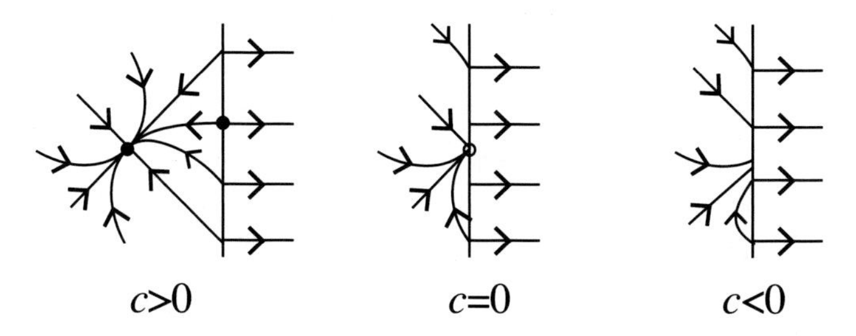

Figure 1.25: A saddle-node bifurcation in a nonsmooth system.

As in the previous example, the two equilibria exist only for $c > 0$, because

- the node exists in the region $x_1 < 0$ since $x_1 = -\frac{c}{4} < 0$ for $c < 0$,
- the saddle leaves the $\lambda \in (-1,+1)$ layer since $\lambda = \frac{c+b-2}{c-b+2} \in (-1,+1)$ for $c < 0$.

As c becomes negative the two equations leave their domains of existence and disappear, in a piecewise-smooth system analogue of the saddle-node bifurcation.

1.6 Multiple switches

An important feature of these methods is that they extend directly to systems that have multiple discontinuities at different thresholds.

1.6.1 Combinations for r switches

Recall that we wrote $\dot{\boldsymbol{x}} = \boldsymbol{f}(\boldsymbol{x};\lambda)$ for a single switch with $\lambda = \text{sign}(\sigma)$. This expression is easy to extend to multiple switches.

Let the switching surface Σ be comprised of m transversally intersecting sub-manifolds, $\Sigma = \Sigma_1 \cup \Sigma_2 \cup \cdots \cup \Sigma_m$, where

$$\begin{aligned} \Sigma_i &= \{\, \boldsymbol{x} \in \mathbb{R}^n : \sigma_i(\boldsymbol{x}) = 0 \,\}, \\ \Sigma &= \{\, \boldsymbol{x} \in \mathbb{R}^n : \sigma(\boldsymbol{x}) = \sigma_1(\boldsymbol{x})\sigma_2(\boldsymbol{x}) \cdots \sigma_m(\boldsymbol{x}) = 0 \,\}, \end{aligned}$$

in terms of smooth scalar functions σ_i. Transversality means the normal vectors $\nabla\sigma_i$ are linearly independent, so the surfaces $\sigma_i = 0$ may touch each other, but never tangentially, see fig. 1.26.

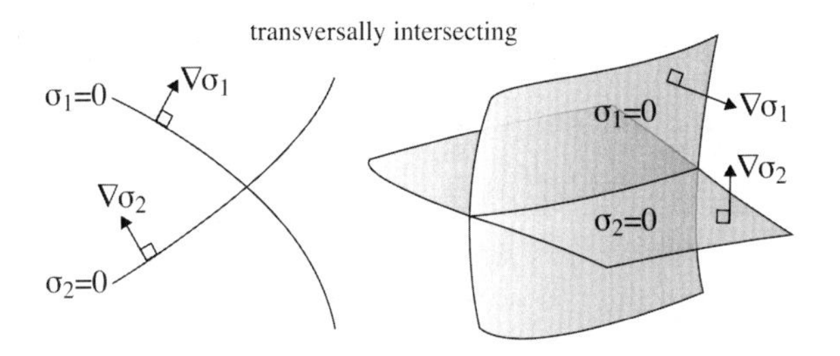

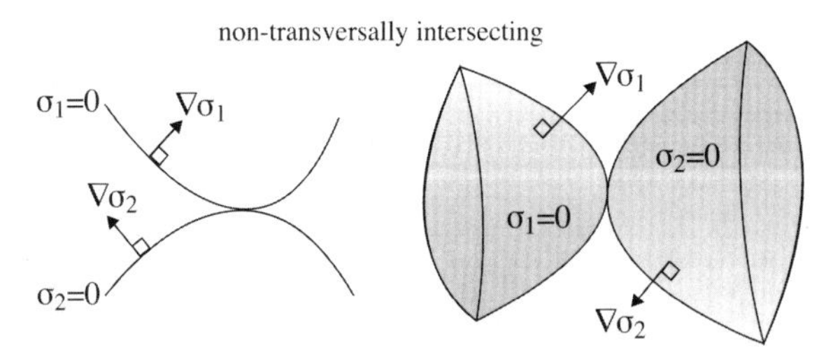

Figure 1.26: In this section we will only consider transversally intersecting surfaces. Tangential intersections occur only under special conditions, they do not persist under typical perturbations, and some of these general methods then need adapting for such atypical scenarios.

The multiplier λ becomes a vector $\boldsymbol{\lambda} = (\lambda_1, \ldots, \lambda_m)$ where each $\lambda_i = \text{sign}(\sigma_i)$. The combination becomes

$$\dot{\boldsymbol{x}} = \boldsymbol{f}(\boldsymbol{x};\boldsymbol{\lambda}) : \quad \begin{array}{ll} \lambda_i = \text{sign}\,(\sigma_i(\boldsymbol{x})) & \text{for } \sigma_i(\boldsymbol{x}) \neq 0, \\ \lambda_i \in (-1,+1) & \text{for } \sigma_i(\boldsymbol{x}) = 0. \end{array} \tag{1.68}$$

For example, we can write $\boldsymbol{f}(\boldsymbol{x};+1,-1,-1,-1) = \boldsymbol{f}^{+---}(\boldsymbol{x})$ and so on, where smooth vector fields $\boldsymbol{f}^i(\boldsymbol{x})$ apply in disjoint regions $\mathcal{R}_i$.

If there are m surfaces $\sigma_1 = 0, \ldots, \sigma_m = 0$, intersecting transversally, these result in 2^m different vector fields $\boldsymbol{f}\boldsymbol{x};\pm_1 1,\ldots,\pm_m 1 = \boldsymbol{f}^{\pm_1\cdots\pm_m}(\boldsymbol{x})$ (on 2^m regions $\mathcal{R}_i$).

1.6.2 Canopy combination

Recall that, for a single switch, the combination was given simply by (1.11) if we assumed linear dependence on λ, or (1.12) if we permitted nonlinear dependence. What does the combination $\boldsymbol{f}(\boldsymbol{x};\boldsymbol{\lambda})$ now look like as a function of $\boldsymbol{x}$, of $\boldsymbol{\lambda} = (\lambda_1,\ldots,\lambda_m)$, and of the constituent vector fields $\boldsymbol{f}^{++\cdots}$, $\boldsymbol{f}^{-+\cdots},\ldots$, (or in decimal indices $\boldsymbol{f}^1$, $\boldsymbol{f}^2$, ...)?

A very general extension of the convex combination to multiple switches is the *convex hull*,

$$\dot{\boldsymbol{x}} = \boldsymbol{f}(\boldsymbol{x};\boldsymbol{\lambda}) = \sum_i \lambda_i \boldsymbol{f}^i(\boldsymbol{x}),$$

summing over all the indices i labelling regions, and subject to a normalization condition $\sum_i \lambda_i = 1$. This means typically that for m switching surfaces we have 2^m vector fields $\boldsymbol{f}^i$, and $2^m - 1$ unknown multipliers λ_i. If we seek sliding motion on those surfaces (for one switch this meant solving $\dot{\sigma} = 0$) we will solve m conditions $\dot{\sigma}_1 = \cdots = \dot{\sigma}_m = 0$. Such a problem is only well posed if the number of unknowns matches the number of conditions, hence if $2^m - 1 = m$, which is only satisfied in the trivial case $m = 1$, i.e., a single switch. The convex hull, therefore, does not give a well-posed expression of the piecewise-smooth system. Fortunately there is a resolution, but like the nonlinear combinations of (1.12), it requires giving up the stipulation of convexity.

Theorem 1.6.1. *If we assume $\dot{\boldsymbol{x}}$ depends multi-linearly on m independent switching multipliers $\lambda_1,\ldots,\lambda_m$, it can be written uniquely as $\dot{\boldsymbol{x}} = \boldsymbol{f}(\boldsymbol{x};\lambda_1,\ldots,\lambda_m)$ using the canopy combination*

$$\dot{\boldsymbol{x}} = \boldsymbol{f}(\boldsymbol{x};\boldsymbol{\lambda}) = \sum_{i_1=\pm}\cdots\sum_{i_m=\pm} \lambda_j^{(i_j)}\lambda_2^{(i_2)}\ldots\lambda_m^{(i_m)}\boldsymbol{f}^{i_1 i_2\cdots i_m}(\boldsymbol{x}), \tag{1.69}$$

using shorthand

$$\lambda_j^{(\pm)} \equiv \tfrac{1}{2}(1 \pm \lambda_j). \tag{1.70}$$

So we can extend the combinations from earlier as the 'canopy' of $\boldsymbol{f}^{\cdots}$ values given by (1.69), introduced in its general form in [60].

We can think of this as a hierarchical application of Filippov's convex combination:

$$\begin{aligned} \boldsymbol{f} &= \tfrac{1}{2}(1+\lambda_1)\boldsymbol{f}^{+\cdots} + \tfrac{1}{2}(1-\lambda_1)\boldsymbol{f}^{-\cdots}, \\ \text{where} \quad \boldsymbol{f}^{i_1\cdots} &= \tfrac{1}{2}(1+\lambda_2)\boldsymbol{f}^{i_1+\cdots} + \tfrac{1}{2}(1-\lambda_2)\boldsymbol{f}^{i_1-\cdots}, \\ \text{where} \quad \boldsymbol{f}^{i_1 i_2\cdots} &= \tfrac{1}{2}(1+\lambda_3)\boldsymbol{f}^{i_1 i_2+\cdots} + \tfrac{1}{2}(1-\lambda_3)\boldsymbol{f}^{i_1 i_2-\cdots}, \\ & \textit{etc.} \end{aligned}$$

where, importantly, the final result is independent of the order in which we build this heirarchy.

For example, for $m=1$, $\dot{\boldsymbol{x}} = \boldsymbol{f}(\boldsymbol{x};\boldsymbol{\lambda}) = \frac{1+\lambda_1}{2}\boldsymbol{f}^+(\boldsymbol{x}) + \frac{1-\lambda_1}{2}\boldsymbol{f}^-(\boldsymbol{x})$ and, for $m=2$, $\dot{\boldsymbol{x}} = \boldsymbol{f}(\boldsymbol{x};\boldsymbol{\lambda}) = \frac{1+\lambda_2}{2}\left\{\frac{1+\lambda_1}{2}\boldsymbol{f}^{++}(\boldsymbol{x}) + \frac{1-\lambda_1}{2}\boldsymbol{f}^{-+}(\boldsymbol{x})\right\} + \frac{1-\lambda_2}{2}\left\{\frac{1+\lambda_1}{2}\boldsymbol{f}^{+-}(\boldsymbol{x}) + \frac{1-\lambda_1}{2}\boldsymbol{f}^{--}(\boldsymbol{x})\right\}$. These are multi-linear in terms of the switching multipliers $\lambda_1, \ldots, \lambda_m$. We can add to this nonlinear dependence on the λ_i's. For a single switch we saw that we could include these via a *hidden* term $(\lambda^2-1)\boldsymbol{g}(\boldsymbol{x};\lambda)$. The principle is the same for multiple switches. We can add a hidden term $\boldsymbol{k}$ which satisfies

$$\sigma_1(\boldsymbol{x})\cdots\sigma_m(\boldsymbol{x})\,\boldsymbol{k}(\boldsymbol{x}) = 0, \tag{1.71}$$

and therefore vanishes from (1.69) outside of the switching surface, where $\boldsymbol{f}(\boldsymbol{x};\boldsymbol{\lambda})$ will reduce to one of the $\boldsymbol{f}^{\pm\cdots\pm}$. This implies that $\boldsymbol{k}$ consists of terms like (λ_i^2-1) $g_i(\boldsymbol{x};\boldsymbol{\lambda})$, which automatically satisfies the orthogonality condition (1.71) for any finite-valued vector fields $\boldsymbol{g}_i$.

1.7 Codimension r switching layers

Taking a general system now in which there are m switches at different surfaces $\sigma_1=0,\ldots,\sigma_m=0$, we will see how to study what happens at a point where $r\leq m$ of these surfaces intersect.

1.7.1 The switching layer for r switches

Extending the method for a single switch, we can magnify each sub-manifold $\sigma_i=0$ into a *layer* over which $\lambda_i\in(-1,+1)$.

Looking at the intersection of r switches in a system with m switches in n dimensions, take coordinates so that $\sigma_j=x_j$ for $j=1,\ldots,r$, with $0<r\leq m\leq n$:

Definition 1.7.1. The **switching layer** on $x_1=x_2=\cdots=x_r=0$ is

$$(\lambda_1,\ldots,\lambda_r,x_{r+1},\ldots,x_n) \in (-1,+1)^r\times\mathbb{R}^{n-1}. \tag{1.72}$$

For each multiplier λ_j we introduce an infinitesimal $\varepsilon_j\to 0$, and extending directly what we had for one switch, for each i we have a (infinitely) fast system $\varepsilon_j\dot{\lambda}_j = \boldsymbol{f}\cdot\nabla\sigma_j$ and $\lambda_j\in(-1,+1)$. Putting this together with the system $\dot{\boldsymbol{x}}=\boldsymbol{f}$ on $\sigma_1=\cdots=\sigma_m=0$, in coordinates where $\sigma_j=x_j$ for $j=1,\ldots,r$, the variation inside the layer is given by an $r+1$ timescale system (made up of times $t,\varepsilon_1 t,\ldots,\varepsilon_r t$):

Definition 1.7.2. The switching layer system inside (1.72) is

$$\begin{aligned} \varepsilon_j \dot{\lambda}_j &= f_j(0,\ldots,0,x_{r+1},\ldots,x_n;\lambda_1,\ldots,\lambda_m), \quad j=1,\ldots,r,\\ \dot{x}_i &= f_i(0,\ldots,0,x_{r+1},\ldots,x_n;\lambda_1,\ldots,\lambda_m), \quad i=r+1,\ldots,n, \end{aligned}$$

in terms of (different) infinitesimals $\varepsilon_j > 0$, in the limit $\varepsilon_j \to 0$.

Associated with these, for local analysis, we have:

Definition 1.7.3. In a system where $\lambda_i = \operatorname{sign}(x_i)$ for $i = 1,\ldots,r$, the switching **layer variable** is the vector

$$\boldsymbol{\xi} = (\xi_1,\xi_2,\ldots,\xi_n) = (\varepsilon_1\lambda_1,\ldots,\varepsilon_r\lambda_r,x_{r+1},\ldots,x_n) \tag{1.73}$$

(i.e., the vector $\boldsymbol{\xi}$ given by replacing x_i by $\varepsilon_i\lambda_i$ in coordinates where $\sigma_i = x_i$ for $i = 1,\ldots,r$).

The fact that the ε_j's need not be the same means, although they all tend to zero, their ratios might not. This can have significance for the dynamics of the layer system. One way to treat it is to let $\varepsilon_j = \kappa_j\varepsilon_1$ for $j = 1,\ldots,r$ with $\kappa_1 = 1$, and then assume that all κ_j are fixed non-vanishing constants, the ratios $\kappa_j = \varepsilon_j/\varepsilon_1$. When we let $\varepsilon_1 \to 0$ then all ε_j will tend to zero, but the constants κ_j will be left behind in our expressions. It then becomes part of the modeling problem, depending on the application for instance, to determine what the values of the κ_j's should be, based on a study of the dynamics they result in, so let's begin seeing what that dynamics might look like.

1.8 Codimension *r* sliding

On the switching surfaces we will have sliding dynamics if $\dot{\sigma}_j = 0$, which implies $\dot{\lambda}_j = 0$ in the switching layer system. So we have:

Definition 1.8.1. The (codimension r) **sliding manifold** is the set of points

$$\mathcal{M} = \left\{ \begin{array}{l} f_j = 0 \text{ on } x_j = 0 \quad \text{for } j = 1,2,\ldots,r \\ \text{with } (\lambda_1,\ldots,\lambda_r) \in (-1,+1)^r \end{array} \right\} \tag{1.74}$$

in the switching layer where

$$0 = f_j(0,\ldots,0,x_{r+1},\ldots,x_n;\lambda_1,\ldots,\lambda_m) \quad \forall\, j=1,\ldots,r$$

on which the sliding dynamics is given by the differential algebraic equations

$$\begin{aligned} 0 &= f_j(0,\ldots,0,x_{r+1},\ldots,x_n;\lambda_1,\ldots,\lambda_m), \quad j=1,\ldots,r,\\ \dot{x}_i &= f_i(0,\ldots,0,x_{r+1},\ldots,x_n;\lambda_1,\ldots,\lambda_m), \quad i=r+1,\ldots,n. \end{aligned} \tag{1.75}$$

This extends the notion of sliding modes to multiple switches. The manifold is invariant where it is normally hyperbolic, where

$$\det \underline{\underline{B}} \neq 0 \quad \text{and} \quad \mathrm{Re}(\text{eigenvalues } \underline{\underline{B}}) \neq 0 \quad \text{where} \quad \underline{\underline{B}} = \left. \frac{\partial(\dot{\lambda}_1, \ldots, \dot{\lambda}_r)}{\partial(\lambda_1, \ldots, \lambda_r)} \right|_{\mathcal{M}} . \tag{1.76}$$

If the eigenvalues of $\underline{\underline{B}}$ have negative real part, then $\mathcal{M}$ is attracting, generalizing the concept of *attracting sliding regions* to the intersections of multiple switches. If none of the eigenvalues have negative real part, then the sliding manifold is repelling, and less likely to play a major role in the local dynamics. If only some of the eigenvalues have negative real part, then the manifold is attracting in some directions and repelling in others.

If we have multi-linear dependence on the λ_j's, as in the general form of the combination (1.69), we may find the algebraic conditions in (1.75) have up to $r!$ solutions (or even more if we have nonlinear dependence on the λ_j's). These correspond to multiple branches of $\mathcal{M}$, meaning multiple independent sliding modes existing at the same $\boldsymbol{x}$ coordinates on Σ, but separated by existing typically at different values of the λ_j's, each with their own attractivity properties.

Example 1.8.2 (Sliding equilibrium)**.** For a system with two switches, take coordinates such that $\sigma_i = x_i$, so $\lambda_i = \operatorname{sign} x_i$, for $i = 1, 2$, and consider vector fields

$$\begin{pmatrix} \dot{x}_1 \\ \dot{x}_2 \end{pmatrix} = \begin{pmatrix} a_1 \\ a_2 \end{pmatrix} + \begin{pmatrix} b_{11} & b_{12} \\ b_{21} & b_{22} \end{pmatrix} \begin{pmatrix} \lambda_1 \\ \lambda_2 \end{pmatrix}$$

where the a_i and b_{ij} may be functions of $\boldsymbol{x}$. There is codimension $r = 1$ (i.e., 'Filippov') sliding:

- on $x_1 = 0$: where $0 = \dot{x}_1 = a_1 + b_{11}\lambda_1 + b_{12} \operatorname{sign} x_2$ gives

$$\lambda_1^\Sigma = -\tfrac{a_1 + b_{12} \operatorname{sign} x_2}{b_{11}} \quad \Rightarrow \quad \dot{x}_2 = a_2 - \tfrac{b_{21}}{b_{11}} (a_1 + b_{12} \operatorname{sign} x_2) + b_{22} \operatorname{sign} x_2, \tag{1.77a}$$

- on $x_2 = 0$: where $0 = \dot{x}_2 = a_2 + b_{21} \operatorname{sign} x_1 + b_{22}\lambda_2$ gives

$$\lambda_2^\Sigma = -\tfrac{a_2 + b_{21} \operatorname{sign} x_1}{b_{22}} \quad \Rightarrow \quad \dot{x}_1 = a_1 - \tfrac{b_{12}}{b_{22}} (a_2 + b_{21} \operatorname{sign} x_1) + b_{11} \operatorname{sign} x_1, \tag{1.77b}$$

  provided λ_1^Σ and/or λ_2^Σ lie in $(-1, +1)$, and otherwise the flow crosses $x_1 = 0$ and $x_2 = 0$ transversally.

The intersection we must treat separately. There is codimension $r = 2$ sliding where

$$\begin{pmatrix} 0 \\ 0 \end{pmatrix} = \begin{pmatrix} \dot{\lambda}_1 \\ \dot{\lambda}_2 \end{pmatrix} = \begin{pmatrix} \dot{x}_1 \\ \dot{x}_2 \end{pmatrix} = \begin{pmatrix} a_1 \\ a_2 \end{pmatrix} + \begin{pmatrix} b_{11} & b_{12} \\ b_{21} & b_{22} \end{pmatrix} \begin{pmatrix} \lambda_1 \\ \lambda_2 \end{pmatrix}$$

giving $\boldsymbol{\lambda}^\Sigma = -\underline{\underline{B}}^{-1} \boldsymbol{a}$, or

$$\begin{pmatrix} \lambda_1^\Sigma \\ \lambda_2^\Sigma \end{pmatrix} = \frac{1}{b_{11}b_{22} - b_{12}b_{21}} \begin{pmatrix} b_{12}a_2 - b_{22}a_1 \\ b_{21}a_1 - b_{11}a_2 \end{pmatrix} \tag{1.78}$$

which exists if it lies in $(-1,+1)^2$.

The attractivity of the sliding mode can be derived either from the eigenvectors and eigenvalues of

$$\frac{d\boldsymbol{f}}{d\boldsymbol{\lambda}} = \begin{pmatrix} b_{11} & b_{12} \\ b_{21} & b_{22} \end{pmatrix},$$

and in simple cases from the directions of the flows along the manifolds $x_1 = 0$ and $x_2 = 0$.

The attractivity of sliding equilibria can be derived from the layer system

$$\begin{pmatrix} \varepsilon_1\dot{\lambda}_1 \\ \varepsilon_2\dot{\lambda}_2 \end{pmatrix} = \begin{pmatrix} a_1 \\ a_2 \end{pmatrix} + \begin{pmatrix} b_{11} & b_{12} \\ b_{21} & b_{22} \end{pmatrix}\begin{pmatrix} \lambda_1 \\ \lambda_2 \end{pmatrix}$$

and considering the eigenvectors and eigenvalues in layer variables

$$\begin{aligned} \begin{pmatrix} \dot{\xi}_1 \\ \dot{\xi}_2 \end{pmatrix} &= \begin{pmatrix} a_1 \\ a_2 \end{pmatrix} + \begin{pmatrix} b_{11} & b_{12} \\ b_{21} & b_{22} \end{pmatrix}\begin{pmatrix} \xi_1/\varepsilon_1 \\ \xi_2/\varepsilon_2 \end{pmatrix} \\ &= \begin{pmatrix} a_1 \\ a_2 \end{pmatrix} + \begin{pmatrix} b_{11} & b_{12} \\ b_{21} & b_{22} \end{pmatrix}\begin{pmatrix} 1/\varepsilon_1 & 0 \\ 0 & 1/\varepsilon_2 \end{pmatrix}\begin{pmatrix} \xi_1 \\ \xi_2 \end{pmatrix} \end{aligned} \tag{1.79}$$

with Jacobian

$$\frac{\partial(\dot{\xi}_1,\dot{\xi}_2)}{\partial(\xi_1,\xi_2)} = \begin{pmatrix} b_{11} & b_{12} \\ b_{21} & b_{22} \end{pmatrix}\begin{pmatrix} 1/\varepsilon_1 & 0 \\ 0 & 1/\varepsilon_2 \end{pmatrix}. \tag{1.80}$$

Example 1.8.3 (Sliding equilibrium in 3D)**.** Eigenvectors of an equilibrium in codimension 2 sliding. Consider the three-dimensional system

$$\begin{pmatrix} \dot{x}_1 \\ \dot{x}_2 \\ \dot{x}_3 \end{pmatrix} = \begin{pmatrix} a_1 \\ a_2 \\ -x_3 \end{pmatrix} + \begin{pmatrix} b_{11} & b_{12} & 0 \\ b_{21} & b_{22} & 0 \\ 0 & 0 & 0 \end{pmatrix}\begin{pmatrix} \lambda_1 \\ \lambda_2 \\ 0 \end{pmatrix}$$

where $\lambda_i = \operatorname{sign} x_i$, for nonzero constants a_i, b_{ij}. The system outside Σ (which is made up of $x_1 = 0$, $x_2 = 0$, and $x_3 = 0$) is non-vanishing, so there are no equilibria outside the switching surface.

Taking the intersection first, the switching layer system is

$$\begin{pmatrix} \varepsilon_1\dot{\lambda}_1 \\ \varepsilon_2\dot{\lambda}_2 \\ \dot{x}_3 \end{pmatrix} = \begin{pmatrix} a_1 + b_{11}\lambda_1 + b_{12}\lambda_2 \\ a_2 + b_{21}\lambda_1 + b_{22}\lambda_2 \\ -x_3 \end{pmatrix} \quad \text{on } x_1 = x_2 = 0. \tag{1.81}$$

In layer variables this is

$$\begin{pmatrix} \dot{\xi}_1 \\ \dot{\xi}_2 \\ \dot{\xi}_3 \end{pmatrix} = \begin{pmatrix} a_1 + b_{11}\xi_1/\varepsilon_1 + b_{12}\xi_2/\varepsilon_2 \\ a_2 + b_{21}\xi_1/\varepsilon_1 + b_{22}\xi_2/\varepsilon_2 \\ -\xi_3 \end{pmatrix} \quad \text{on } x_1 = x_2 = 0. \tag{1.82}$$

This has a unique equilibrium at

$$(\lambda_1, \lambda_2, x_3) = \frac{(a_2 b_{12} - a_1 b_{22}, a_1 b_{21} - a_2 b_{11}, 0)}{b_{11} b_{22} - b_{12} b_{21}}, \tag{1.83}$$

the first two components of which must lie inside $(-1, +1)$, otherwise the equilibrium ceases to exist.

Where does it go if it ceases to exist? At least in smooth systems, one equilibrium cannot simply vanish without colliding and annihilating with another, and the same proves to be true for nonsmooth systems. We return to this question in the next section.

The layer Jacobian of the equilibrium is

$$\underline{\underline{J}} = \begin{pmatrix} \varepsilon_1^{-1} b_{11} & \varepsilon_2^{-1} b_{12} & 0 \\ \varepsilon_1^{-1} b_{21} & \varepsilon_2^{-1} b_{22} & 0 \\ 0 & 0 & -1 \end{pmatrix} = \begin{pmatrix} b_{11} & b_{12} & 0 \\ b_{21} & b_{22} & 0 \\ 0 & 0 & -1 \end{pmatrix} \begin{pmatrix} 1/\varepsilon_1 & 0 & 0 \\ 0 & 1/\varepsilon_2 & 0 \\ 0 & 0 & 1 \end{pmatrix} \tag{1.84}$$

with eigenvalues and corresponding eigenvectors

$$\begin{aligned} \nu_1 &= \left(\tfrac{b_{11}+\kappa b_{22}}{2} + \sqrt{R}\right)/\varepsilon_1 : & \boldsymbol{v}_1 &= (\nu_1 - \kappa b_{22}, b_{21}, 0)^{\mathsf{T}}, \\ \nu_2 &= \left(\tfrac{b_{11}+\kappa b_{22}}{2} - \sqrt{R}\right)/\varepsilon_2 : & \boldsymbol{v}_2 &= (\nu_2 - \kappa b_{22}, b_{21}, 0)^{\mathsf{T}}, \\ \nu_3 &= -1 : & \boldsymbol{v}_3 &= (0, 0, 1)^{\mathsf{T}}, \end{aligned} \tag{1.85}$$

where $\kappa = \varepsilon_1/\varepsilon_2$ and $R = \left(\frac{b_{11}+\kappa b_{22}}{2}\right)^2 + \kappa b_{12} b_{21} - \kappa b_{11} b_{22}$.

The eigenvector ν_3 gives a finite rate of attraction along the intersection.

The eigenvalues $\nu_{1,2}$ have a magnitude that is infinite as $\varepsilon_{1,2} \to 0$, since they describe dynamics in the plane transverse to the intersection, i.e., in the directions out of the codimension 2 sliding region. Their eigenvectors, however, are finite, assuming that the ratio $\kappa = \varepsilon_1/\varepsilon_2$ is finite and nonzero as $\varepsilon_{1,2} \to 0$.

Thus the parameters b_{ij} and the ratio κ determine whether the equilibrium is a node, focus, or saddle in the $\lambda_{1,2}$ plane. This closely mirrors what happens outside the intersection, and if $\kappa = 1$ it corresponds directly. Note, however, that the trace of $\underline{\underline{J}}$,

$$\operatorname{Tr} \underline{\underline{J}} = \frac{b_{11} + \kappa b_{22}}{\varepsilon_1}, \tag{1.86}$$

depends on the ratio $\kappa = \varepsilon_1/\varepsilon_2$ significantly — it changes sign if $\kappa = -b_{11}/b_{22}$, meaning that the stability of the sliding equilibrium can change depending on the ratio κ. (Certain consequences of this in the context of simulating nonsmooth systems by smoothing the discontinuity can be found in [55].)

Take some values of the a_i's, B_{ij}'s, and κ and try it out for yourself.

1.9 Boundary equilibrium bifurcations

Recall that the layer system is defined only on $\boldsymbol{\lambda} \in (-1, +1)^r$ (for r switches). If any of the λ_j's satisfying $\boldsymbol{f}_j = 0$ ($\forall\, j = 1, \ldots, r$) sits at $\lambda_j^{\Sigma} = \pm 1$, then the point lies on a boundary of codimension r sliding.

Example 1.9.1 (Boundary equilibrium bifurcation between codimension 1 and 2 sliding)**.** Consider again the three-dimensional system above. There are no equilibria outside the switching surface, and we saw that an equilibrium exists inside the intersection $x_1 = x_2 = 0$ at

$$(\lambda_1, \lambda_2, x_3) = \frac{(a_2 b_{12} - a_1 b_{22}, a_1 b_{21} - a_2 b_{11}, 0)}{b_{11} b_{22} - b_{12} b_{21}}, \tag{1.87}$$

only for

$$\left| \frac{a_2 b_{12} - a_1 b_{22}}{b_{11} b_{22} - b_{12} b_{21}} \right| < 1, \quad \left| \frac{a_1 b_{21} - a_2 b_{11}}{b_{11} b_{22} - b_{12} b_{21}} \right| < 1. \tag{1.88}$$

When the parameters lie outside this region, the equilibrium inside the intersection no longer exists. Bifurcations occur at

$$\left| \frac{a_2 b_{12} - a_1 b_{22}}{b_{11} b_{22} - b_{12} b_{21}} \right| = 1 \quad \text{and} \quad \left| \frac{a_1 b_{21} - a_2 b_{11}}{b_{11} b_{22} - b_{12} b_{21}} \right| = 1. \tag{1.89}$$

Where does the equilibrium go?

There are sliding regions on the switching manifolds $x_1 = 0$ or $x_2 = 0$ outside the intersection. The bifurcation that actually occurs at these values will be degenerate, with the entire sliding vector field along $x_2 = 0$ and $x_1 = 0$ vanishing respectively at these two parameter values. A more generic system is easily obtained, say by perturbing this system to

$$\begin{pmatrix} \dot{x}_1 \\ \dot{x}_2 \\ \dot{x}_3 \end{pmatrix} = \begin{pmatrix} a_1 + c_1 x_1 \\ a_2 + c_2 x_2 \\ -x_3 \end{pmatrix} + \begin{pmatrix} b_{11} & b_{12} & 0 \\ b_{21} & b_{22} & 0 \\ 0 & 0 & 0 \end{pmatrix} \begin{pmatrix} \lambda_1 \\ \lambda_2 \\ 0 \end{pmatrix}$$

which does not change the system on the intersection, hence our analysis so far stands.

On the switching manifolds $x_1 = 0$ or $x_2 = 0$ outside the intersection, the layer systems are

$$\begin{aligned} \begin{pmatrix} \varepsilon_1 \dot{\lambda}_1 \\ \dot{x}_2 \\ \dot{x}_3 \end{pmatrix} &= \begin{pmatrix} a_1 + b_{11}\lambda_1 + b_{12} \operatorname{sign}(x_2) \\ a_2 + c_2 x_2 + b_{21}\lambda_1 + b_{22} \operatorname{sign}(x_2) \\ -x_3 \end{pmatrix} \quad \text{on } x_1 = 0 \neq x_2, \\ \begin{pmatrix} \dot{x}_1 \\ \varepsilon_2 \dot{\lambda}_2 \\ \dot{x}_3 \end{pmatrix} &= \begin{pmatrix} a_1 + c_1 x_1 + b_{11} \operatorname{sign}(x_1) + b_{12}\lambda_2 \\ a_2 + b_{21} \operatorname{sign}(x_1) + b_{22}\lambda_2 \\ -x_3 \end{pmatrix} \quad \text{on } x_2 = 0 \neq x_1. \end{aligned} \tag{1.90}$$

Sliding (where the $\dot{\lambda}_i$ subsystems vanish) occurs for $\lambda_1^\Sigma = -\frac{a_1 + b_{12}\operatorname{sign}(x_2)}{b_{11}}$ and $\lambda_2^\Sigma = -\frac{a_2 + b_{21}\operatorname{sign}(x_1)}{b_{22}}$ respectively, giving dynamics

$$\begin{aligned} \dot{x}_2 &= c_2 x_2 + \tfrac{a_2 b_{11} - a_1 b_{21}}{b_{11}} - \tfrac{b_{21} b_{12} - b_{11} b_{22}}{b_{11}} \operatorname{sign}(x_2) \quad \text{on } x_1 = 0 \neq x_2, \\ \dot{x}_1 &= c_1 x_1 + \tfrac{a_1 b_{22} - a_2 b_{12}}{b_{22}} - \tfrac{b_{12} b_{21} - b_{11} b_{22}}{b_{22}} \operatorname{sign}(x_1) \quad \text{on } x_2 = 0 \neq x_1. \end{aligned} \tag{1.91}$$

We can see from these that, at the bifurcation values, as the equilibrium vanishes from codimension two sliding on the intersection, it either passes into one of these two systems, or collides with another equilibrium and the two annihilate in another example of a discontinuity-induced saddlenode bifurcation. We leave it as an exercise to explore the different scenarios.

1.10 Stability, equivalence, and bifurcation

The notions of equivalence between systems, and structural stability of a system within a given class, are incredibly important in smooth systems, and are quite easy to extend to piecewise-smooth systems, but use them with care.[3]

There are three notions of equivalence between systems that are often useful.

Definition 1.10.1. Orbital, differentiable, and topological equivalence:

(i) **Orbital equivalence**: If two vector fields $\boldsymbol{f}$ and $\hat{\boldsymbol{f}}$ are related by $\boldsymbol{f}(\boldsymbol{x};\boldsymbol{\lambda}) = \mu(\boldsymbol{x};\boldsymbol{\lambda})\hat{\boldsymbol{f}}(\boldsymbol{x};\boldsymbol{\lambda})$ for some continuous positive definite scalar function $\mu(\boldsymbol{x};\boldsymbol{\lambda})$, the orbits of the systems $\dot{\boldsymbol{x}} = \boldsymbol{f}(\boldsymbol{x};\boldsymbol{\lambda})$ and $\dot{\boldsymbol{x}} = \hat{\boldsymbol{f}}(\boldsymbol{x};\boldsymbol{\lambda})$ are identical up to a time rescaling. Their phase portraits are then identical.

(ii) **q-Conjugacy**: If two vector fields $\boldsymbol{f}$ and $\hat{\boldsymbol{f}}$ are related by a q-times differentiable mapping $\boldsymbol{h}$ which takes orbits of $\dot{\boldsymbol{x}} = \boldsymbol{f}(\boldsymbol{x};\boldsymbol{\lambda})$ to those of $\dot{\boldsymbol{x}} = \hat{\boldsymbol{f}}(\boldsymbol{x};\boldsymbol{\lambda})$, preserving direction but not necessarily scaling of time, then the vector fields $\boldsymbol{f}(\boldsymbol{x};\boldsymbol{\lambda})$ and $\hat{\boldsymbol{f}}(\boldsymbol{x};\boldsymbol{\lambda})$ are said to be q-conjugate (or q-differentiably equivalent, sometimes called C^q equivalence).

(iii) **Topological equivalence**: If $q = 0$ in (ii) and the switching surface is preserved, the vector fields $\boldsymbol{f}(\boldsymbol{x};\boldsymbol{\lambda})$ and $\hat{\boldsymbol{f}}(\boldsymbol{x};\boldsymbol{\lambda})$ are said to be *topologically equivalent*. That is, topological equivalence between the vector fields $\boldsymbol{f}(\boldsymbol{x};\boldsymbol{\lambda})$ and $\hat{\boldsymbol{f}}(\boldsymbol{x};\boldsymbol{\lambda})$ means that $\boldsymbol{f}$ and $\hat{\boldsymbol{f}}$ are related by a continuous mapping $\boldsymbol{h}$ which takes orbits of $\dot{\boldsymbol{x}} = \boldsymbol{f}(\boldsymbol{x};\boldsymbol{\lambda})$ to those of $\dot{\boldsymbol{x}} = \hat{\boldsymbol{f}}(\boldsymbol{x};\boldsymbol{\lambda})$, preserving direction but not necessarily scaling of time, and maps the switching surface of one system to that of the other preserving orientation with respect to orbits.

It is important to include the switching surface explicitly in the definition of topological equivalence, otherwise, for example, a system that crosses a switching surface is equivalent to a smooth system with no switching surface at all.

A q-conjugacy with:

- $q = 0$ preserves spatial and temporal topology of orbits themselves, but does not preserve eigenvalues,
- $q > 1$ preserves only the relative sizes of eigenvalues,

[3] And when 'taking care', we should all beware of the usage of the notion of 'normal forms' in nonsmooth systems. This is a very powerful concept from smooth systems that is not yet well understood in nonsmooth systems. This can lead to confusion, because the strong associations of normal forms with generality and universality in smooth systems have not been extended to nonsmooth systems in many of the cases where they are employed.

- q infinite preserves the eigenvalues associated with any equilibria.

So a node and a focus are topologically equivalent if they have the same attractivity, fig. 1.27, but only on a region that does not include the switching surface.

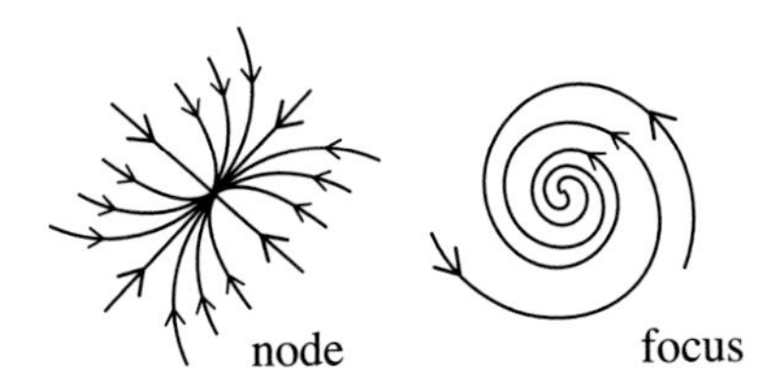

Figure 1.27: A node and focus are topologically equivalent.

At a switching surface, a node and a focus of the same attractivity are not topologically equivalent, fig. 1.28, because all orbits hit the switching surface in both forward and backward time around a boundary focus, while some orbits (shaded in figure) contact the switching surface only in one direction near a boundary node.

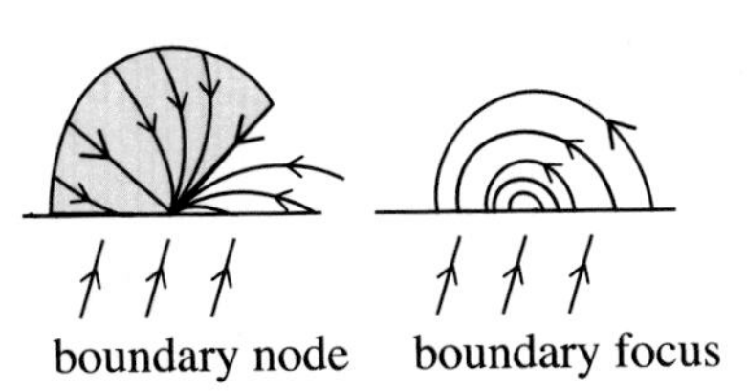

Figure 1.28: A node and focus at a switching surface are not topologically equivalent. (We may also call these a boundary-node and boundary-focus).

Note that systems may be equivalent but have very different functional expressions, and conversely, two systems that appear similar in their functional expressions may in fact not be equivalent. That is why these technical definitions of equivalence are so important.

A system is considered robust in its behaviour — or *structurally stable* — if small changes in its expression produce equivalent systems. Intuitively, a small perturbation involves the addition of a small term in the equations, but does not qualitatively alter the dynamics.

The precise definitions (see, e.g., Guckenheimer & Holmes 2002) can be extended (see, e.g., [29]) formally, for example:

Definition 1.10.2. If $\boldsymbol{f} \in \mathbb{R}^n$ and $\sigma \in \mathbb{R}$ are r times partially differentiable (vector and scalar valued resp.) functions, in which the level set $\sigma = 0$ is the switching surface Σ of $\boldsymbol{f}$, then for some $0 < q \leq r$ and $\varepsilon > 0$, the functions $\tilde{\boldsymbol{f}}$ and $\tilde{\sigma}$ are a perturbation of f and σ of size ε, of differentiability class q, if there is a compact set $K \subset \mathbb{R}^{n+1}$ such that $\boldsymbol{f} = \tilde{\boldsymbol{f}}$ and $\sigma = \tilde{\sigma}$ on the complement set $K_c = \mathbb{R}^{n+1} - K$ and for all $i_1, i_2, \ldots, i_n$, with $i = i_1 + i_2 + \cdots + i_n \leq q$ we have $\left|\left(\partial^i/\partial x_1^{i_1} \cdots \partial x_n^{i_n}\right)(\boldsymbol{f} - \tilde{\boldsymbol{f}})\right| < \varepsilon$ and $\left|\left(\partial^i/\partial x_1^{i_1} \cdots \partial x_n^{i_n}\right)(\sigma - \tilde{\sigma})\right| < \varepsilon$.

This is a useful definition, intuitive if technical looking, so it is more important to understand what it means. Perturbing a piecewise-smooth systems is a subtle act, but it is vital to our notion of 'robustness' of a system, or structural stability.

Definition 1.10.3. A vector field $\boldsymbol{f} \in \mathbb{R}^n$ is **structurally stable** if there is an $\varepsilon > 0$ such that all differentiable ($q = 1$) order ε perturbations of $\boldsymbol{f}$ are topologically equivalent to $\boldsymbol{f}$.

In a smooth system we typically consider only perturbations that are continuous or differentiable in $\boldsymbol{x}$. We could insist on the same in piecewise-smooth systems, and this will ensure that sliding is preserved, for example, but demanding only perutbrations with some level of smoothness in a *non*smooth system seems a little *too* safe. The way we have set things up in terms of *combinations*, we can actually do somewhat better.

If we write

$$\dot{\boldsymbol{x}} = \boldsymbol{f}^+(\boldsymbol{x}) \text{ on } \mathcal{R}_+, \quad \dot{\boldsymbol{x}} = \boldsymbol{f}^-(\boldsymbol{x}) \text{ on } \mathcal{R}_-, \quad \ldots, \tag{1.92}$$

then can we perturb in one region (say $\dot{\boldsymbol{x}} = \boldsymbol{f}^+(\boldsymbol{x}) + \mu$ for small μ) but not another?

This means introducing a perturbation of μ in $\mathcal{R}_1$ but no perturbation elsewhere, i.e., the perturbation is discontinuous.

Normally we do not allow discontinuous perturbations. Even a simple smooth system with a stable equilibrium is not structurally stable under discontinuous perturbations. For example $\dot{x}_1 = x_1$ is equivalent to $\dot{x}_1 = x_1 - \mu$ (the equilibrium is slightly shifted), but not to $\dot{x}_1 = x_1 - \mu \operatorname{sign} x_1$ (the equilibrium splits into three, equilibria at $x_1 = \pm\mu$ and a sliding equilibrium at $x_1 = 0$!), even though these all tend to the same thing as $\mu \to 0$.

Even if we disallow such obviously absurd perturbations, what conditions must we place on the perturbation at the switching surface to make sure sliding dynamics is preserved? You will find partial answers to these in [21, 29], which seem to suggest only differentiable perturbations should be allowed — we cannot perturb $\boldsymbol{f}^+$ without perturbing $\boldsymbol{f}^-$ the same amount.

Because we have expressed our system in the form $\dot{\boldsymbol{x}} = \boldsymbol{f}(\boldsymbol{x}; \boldsymbol{\lambda})$, however, we can do a little more, and allow perturbations that are at least partially differentiable in $\boldsymbol{x}$ or $\boldsymbol{\lambda}$.

Intuitively, small changes in the dependence on $\boldsymbol{\lambda}$ will make only small changes in the equations for sliding, for example, and preserve equivalence. This is despite the fact that adding, say, $\mu\boldsymbol{\lambda}$, means adding a perturbation that is different in different regions. The discontinuity of the perturbation with respect to $\boldsymbol{x}$ is implicitly hidden inside $\boldsymbol{\lambda}$.

So if we want to perturb $\dot{x} = x$ with respect to $\lambda = \operatorname{sign} x$, we must first consider this to be a system $\dot{x} = f(x; \lambda)$, i.e., so that the switching surface $x = 0$ is part of the system's definition (even if it doesn't cause a jump in the vector field).

We can now see immediately that the system $\dot{x} = x$ with $\lambda = \text{sign}(x)$ is structurally unstable, because x lies on the switching surface, and any small perturbation will kick it off, giving a non-equivalent system. The switching surface matters even if we cannot see it in the vector field itself! This is because a switching surface/multiplier is defined as part of the system.

Example 1.10.4. 1. The system defined as $\dot{x} = x$ for $x \in \mathbb{R}$ is structurally stable as it has an equilibrium on the distinguished point $x = 0$, but the system defined as

$$\dot{x} = x \quad \text{for } x \gtrless 0 \tag{1.93}$$

about a switching surface $\Sigma = \{x = 0\}$ is not structurally stable. A perturbation $\dot{x} = \mu + x$ or $\dot{x} = \mu\lambda + x$ when $\lambda = \text{sign}(x)$ for small μ, creates a non-equivalent system, where the equilibrium (or equilibria) do not lie on the distinguished point $x = 0$.

2. The system

$$\dot{x} = x - 1 \quad \text{with } \lambda = \operatorname{sign} x \tag{1.94}$$

is structurally stable. If we perturb to $\dot{x} = x - 1 - \mu\lambda$ for small μ, the resulting system is equivalent.

3. The system

$$\dot{x} = -\lambda \quad \text{with } \lambda = \operatorname{sign} x$$

is structurally stable.

Both Filippov [29] and Teixeira [93] consider pseudo-orbits to be a determining factor in the equivalence of systems. It depends what you want to study about a system, but I prefer to exclude them. A pseudo-orbit is a concatenation of trajectories that does not preserve the direction of time, therefore its topological existence seems to be of no dynamical (or as far as we know physical) significance, though it would be interesting if examples were found where this proved not to be the case.

Example 1.10.5 (Stable or not?).

- A fused centre is not structurally stable. It is comprised entirely of closed orbits that cross through a switching surface, formed by the piecewise-smooth fusing of two parabolic systems.
- Perturbation of a fused centre results in a (repelling or attracting) fused focus, which typically is structurally stable, possibly surrounded by one or more isolated closed orbits.
- A pseudo-fused centre (similar to the centre but the orbits travel the 'wrong way' on one side of the surface) is typically structurally stable, and is dynamically no different from a ...
- ... pseudo-fused focus.

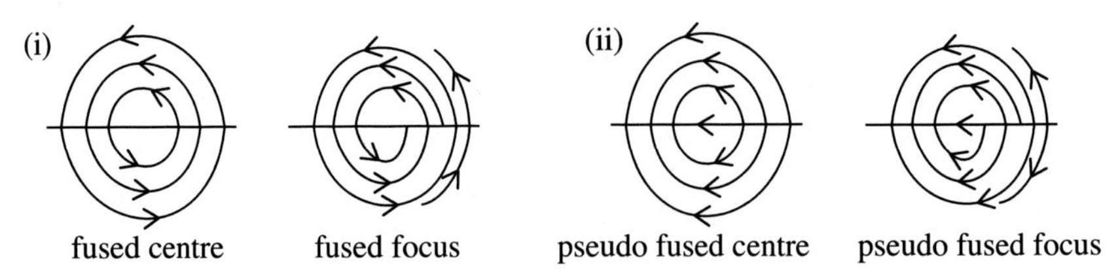

Figure 1.29: (i) The fused centre (where every orbit is a closed cycle) and fused focus (where every orbit spirals into a point) are not equivalent; the centre is structurally unstable while the focus is stable. (ii) Whether the same is true for for the corresponding pseudo-singularities is a matter of interpretation, but all trajectories slide from the right, may pass through the upper or lower half plane, before sliding again to the left.

The systems we will study generally depend on variables $\boldsymbol{x} = (x_1, x_2, \ldots, x_n)$, switching multipliers $\boldsymbol{\lambda} = (\lambda_1, \lambda_2, \ldots, \lambda_m)$, and parameters $\boldsymbol{p} = (a, b, c, \ldots)$. Structurally unstable systems may occur at particular values of the parameters.

Definition 1.10.6. A **bifurcation set** is the set of parameters $\boldsymbol{p} = (a, b, c, \ldots)$ for which the system $\dot{\boldsymbol{x}} = \boldsymbol{f}(\boldsymbol{x}; \boldsymbol{\lambda}; \boldsymbol{p})$ is structurally unstable. Any point $\boldsymbol{x}$ in a neighbourhood of which the system is structurally unstable is a **singularity** (or singular point).

A *bifurcation* of the system is a qualitative change that takes place as we vary parameters through the bifurcation set.

Bifurcations can take place in a region where $\dot{\boldsymbol{x}} = \boldsymbol{f}^i$ is smoothly varying, which may be outside the switching surface, or may be inside the surface if $\boldsymbol{f}^i$ is one of the sliding vector fields. These are covered by the bifurcation theory of smooth dynamical systems.

Then there are a whole new array of bifurcations that cannot occur in smooth dynamical systems, because they involve the discontinuity in a nontrivial way.

Definition 1.10.7. A bifurcation is said to be **discontinuity-induced** in a system $\dot{\boldsymbol{x}} = \boldsymbol{f}(\boldsymbol{x}; \boldsymbol{\lambda})$ if it involves a singular point on the boundary of a sliding region.

This tightens a definition given in [21], which would permit any bifurcation in a discontinuous system to be considered discontinuity-induced. To our knowledge at present, all bifurcations that involve the discontinuity in a nontrivial way appear to involve the boundary of a sliding region. Still, the definition almost certainly is still not perfect.

1.11 Discontinuity-induced phenomena

We end this course with a brief survey of some novel phenomena this course provides the foundations to explore. Many of these are the starting points for open problems.

1.11.1 Local bifurcation points: Tangencies

Above we looked a tangencies as the boundaries of sliding regions. We also defined a *discontinuity-induced bifurcation* as one occurring at such a boundary. This means that tangencies lie at the heart of most of the interesting phenomena in piecewise-smooth flows.

Tangencies are a rich source of bifurcations. Already in a planar system they have a number of cases, shown in the table in fig. 1.30 on the next page.

One can derive the fold cases shown from:

$$\dot{\boldsymbol{x}} = \boldsymbol{f}(\boldsymbol{x};\lambda) = \tfrac{1}{2}(1+\lambda)\begin{pmatrix} a(x_2-1)+bx_1 \\ -1 \end{pmatrix} + \tfrac{1}{2}(1-\lambda)\begin{pmatrix} \alpha_1+\alpha_2 x_2 \\ a \end{pmatrix} \tag{1.95}$$

and the cusp cases shown from:

$$\dot{\boldsymbol{x}} = \boldsymbol{f}(\boldsymbol{x};\lambda) = \tfrac{1}{2}(1+\lambda)\begin{pmatrix} x_2^2+b \\ \pm 1 \end{pmatrix} + \tfrac{1}{2}(1-\lambda)\begin{pmatrix} 1 \\ 0 \end{pmatrix} \tag{1.96}$$

with $\lambda = \text{sign}(x_1)$, where a, b, α_i, are constants. The system with $b = 0$ is structurally unstable, and so a bifurcation takes place as b changes sign. These are therefore *one-parameter* (or *codimension one*) bifurcations.

The simplest tangency between the upper vector field and the switching surface occurs at a point where $\boldsymbol{f}^+ \cdot \nabla\sigma = 0$, but $(\boldsymbol{f}^+ \cdot \nabla)^2\sigma \neq 0$, implying that $\boldsymbol{f}^+ \cdot \nabla\sigma$ will be nonzero nearby. Then either $\boldsymbol{f}^+ \cdot \nabla\sigma > 0$ so the flow is away from the surface in $\sigma > 0$, or $\boldsymbol{f}^+ \cdot \nabla\sigma < 0$ so the flow is towards the surface in $\sigma > 0$.

Similarly the simplest tangency between the lower vector field and the switching surface occurs at a point where $\boldsymbol{f}^- \cdot \nabla\sigma = 0$, but $(\boldsymbol{f}^- \cdot \nabla)^2\sigma \neq 0$, implying that $\boldsymbol{f}^- \cdot \nabla\sigma$ will be nonzero nearby. Then either $\boldsymbol{f}^- \cdot \nabla\sigma > 0$ so the flow is towards the surface in $\sigma < 0$, or $\boldsymbol{f}^- \cdot \nabla\sigma < 0$ so the flow is away from the surface in $\sigma < 0$.

We call these simplest tangencies *folds* — the flow folds parabolically away from the switching surface. The point where the fold occurs for the upper and lower vector fields will not typically be the same. When they *are* the same, the system is not structurally stable, a small perturbation will push them to different points, resulting in a bifurcation. A study of the different cases can be found in [29, 67], we shall give only a brief description.

First, the flows may curve towards the surface on both sides, forming *invisible* fold-folds, they may curve away from the surface on both sides, forming *visible* fold-folds, or they may be a mixture of the two. Within these, the flows above and below the surface may both point to the left (both pointing to the right is just a reflection), or one may point to the left and the other to the right. For the visible fold-fold and invisible fold-fold these give a complete classification. A bifurcation occurs as the folds exchange relative position on the switching surface. Some further analysis, of the kind shown in earlier sections, reveals the existence of sliding equilibria. Closer thought about the geometry of the first invisble fold-fold case also reveals that a limit cycle must exist on one side of the bifurcation.

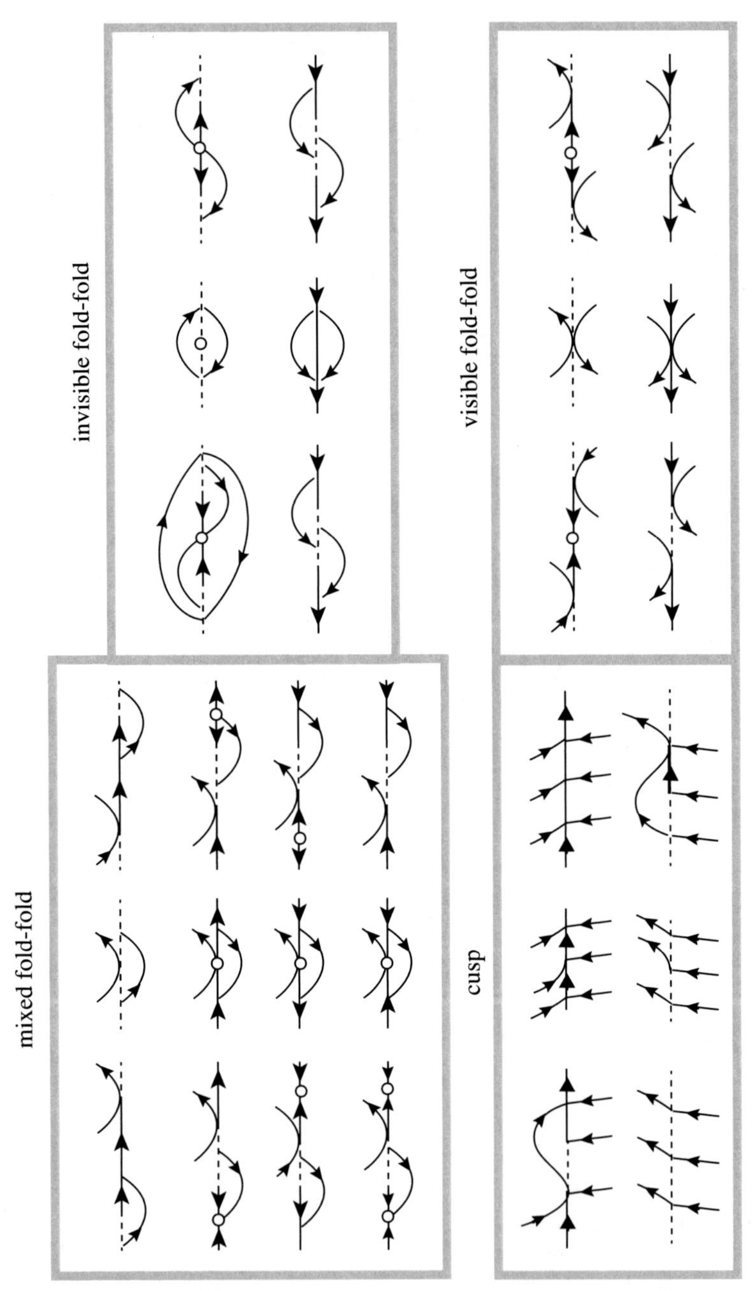

Figure 1.30: Codimension 1 tangencies and their unfoldings.

For the mixed fold-folds things are complicated by the sliding dynamics. In the case where one flow is pointing to the left and one to the right, carrying out the analysis we have shown in earlier sections, we find that the sliding flow may contain a node which changes attractivity as it transfers between branches of attracting and repelling sliding, or a saddle which transfers between branches of attracting and repelling sliding, or a saddle and node which annihilate.

The next highest order of tangency is a cusp, where the flow forms a cubic or cusp-like intersection with the switching surface. A perturbation splits this into a pair of folds, one visible and one invisible. The cases to consider then involve only whether a region of sliding opens up between the folds (with crossing elsewhere), or a region of crossing opens up between the folds (with sliding elsewhere). The flow below the switching surface is assumed to be simple.

Any other case of these fold-folds or cusps is either equivalent to those in fig. 1.30, or else involves a higher degree of bifurcation that requires more than one parameter to unfold it fully.

These become much richer still in three (or more) dimensions, of course, see for example the ongoing work of Teixeira starting with [93], but in three or more dimensions much remains to be discovered.

One particular singularity has caused much confusion and misunderstanding over the last 30 years, yet in principle it could hardly be more simple. If we take the fold-fold cases from the planar classification in fig. 1.30, and add a dimension, each fold (tangency) occurs along a line rather than at a point, and typically these will cross at a point on the switching surface, creating a generic singularity.

Example 1.11.1 (The two-fold singularity). Consider the piecewise linear system

$$(\dot{x}_1, \dot{x}_2, \dot{x}_3) = \begin{cases} (-x_2, a, v) & \text{if } x_1 > 0, \\ (\ \ x_3, w, b) & \text{if } x_1 < 0. \end{cases} \tag{1.97}$$

Exercise: see what you can find out about this. A short article [64] summarizes what we understand about it, and the convoluted history that brought us here. There are three main 'flavours' depending on whether each fold is visible or invisible. Then there are lots of sub-cases depending on the sliding dynamics, determined by v and w.

Tangencies are much more than singularities in their own right. Any object — any attractor or orbit — that acquires or loses a connection with the switching surface must do so via a tangency. We will see a few examples below.

1.11.2 Local bifurcation points: Boundary equilibria

Definition 1.11.2. An equilibrium of a vector field $\boldsymbol{f}^i$ lying on the switching surface (or of a codimension r sliding vector field lying on a codimension $r+1$ switching intersection) is a **boundary equilibrium**.

A boundary equilibrium is structurally unstable, and is typically a bifurcation point.

A prototype for boundary equilibria is quite easy to write down. We take a general equilibrium on one side of the switching surface, and a constant but *sufficiently general* vector field on the other side. (The term in italics turns out to be important). For example we can write

$$\dot{\boldsymbol{x}} = \boldsymbol{f}(\boldsymbol{x};\lambda) = \tfrac{1}{2}(1+\lambda)\underline{\underline{A}}.\begin{pmatrix} x_1-\mu \\ x_2 \\ x_3 \\ \vdots \end{pmatrix} + \tfrac{1}{2}(1-\lambda)\begin{pmatrix} 1 \\ d_1 \\ d_2 \\ \vdots \end{pmatrix} \tag{1.98}$$

with $\lambda = \text{sign}(x_1)$, where the $n \times n$ matrix $\underline{\underline{A}}$ and n-vector $(1, d_1, \ldots, d_{n-1})$ are constants.

For a planar system the full classification of one-parameter boundary equlibrium bifurcations is shown in fig. 1.31. This can be recreated using the prototype above for two dimensions, and in fact it is sufficient to take, with $\lambda = \text{sign}(x_1)$,

$$\dot{\boldsymbol{x}} = \boldsymbol{f}(\boldsymbol{x};\lambda) = \tfrac{1}{2}(1+\lambda)\begin{pmatrix} a(x_1-\mu)+bx_2 \\ cx_2 \end{pmatrix} + \tfrac{1}{2}(1-\lambda)\begin{pmatrix} 1 \\ d \end{pmatrix}. \tag{1.99}$$

A boundary equilibrium occurs at $\mu = 0$, when the equilibrium of the $\lambda = +1$ system lies at $(x_1, x_2) = (0,0)$ on the switching surface $x_1 = 0$.

As we change μ, a *boundary equilibrium bifurcation* occurs. For $\mu > 0$ the equilibrium lies in $x_1 > 0$. For $\mu < 0$ this equilibrium no longer exists in $x_1 > 0$. Applying the methods from earlier sections, we can show either that the equilibrium becomes a sliding equilibrium, or it collides with a coexisting sliding equilibrium in a nonsmooth saddle-node bifurcation.

I promised you that tangencies lay at the heart of discontinuity-induced bifurcations. The tangency in the system above can only occur in the $\lambda = +1$ system (hopefully you can see why very easily), so it lies where

$$x_1 = \dot{x}_1 = 0 \ \& \ \lambda = +1 \quad \Rightarrow \quad a(-\mu) + bx_2 = 0 \quad \Rightarrow \quad x_2 = \mu a/b. \tag{1.100}$$

Is this a visible or invisible fold? A visible tangency curves away from the surface, an invisible curves towards it. In $x_1 \geq 0$ this means $\ddot{x}_1 > 0$ for visible or $\ddot{x}_1 < 0$ for invisible (in $x_1 \leq 0$ the conditions are means $\ddot{x}_1 < 0$ for visible or $\ddot{x}_1 > 0$ for invisible), so evaluate

$$\ddot{x}_1 = a\dot{x}_1 + b\dot{x}_2 = a\left\{a(x_1-\mu)+bx_2\right\} + bcx_2 = \mu a^2 c. \tag{1.101}$$

The tangency therefore switches between visible and invisible as μ changes sign, i.e., as the bifurcation 'unfolds', and as the equilibrium contacts the boundary. The curvature $\ddot{x}_1$ vanishes at $\mu = 0$, so at the bifurcation point itself the tangency is degenerate.

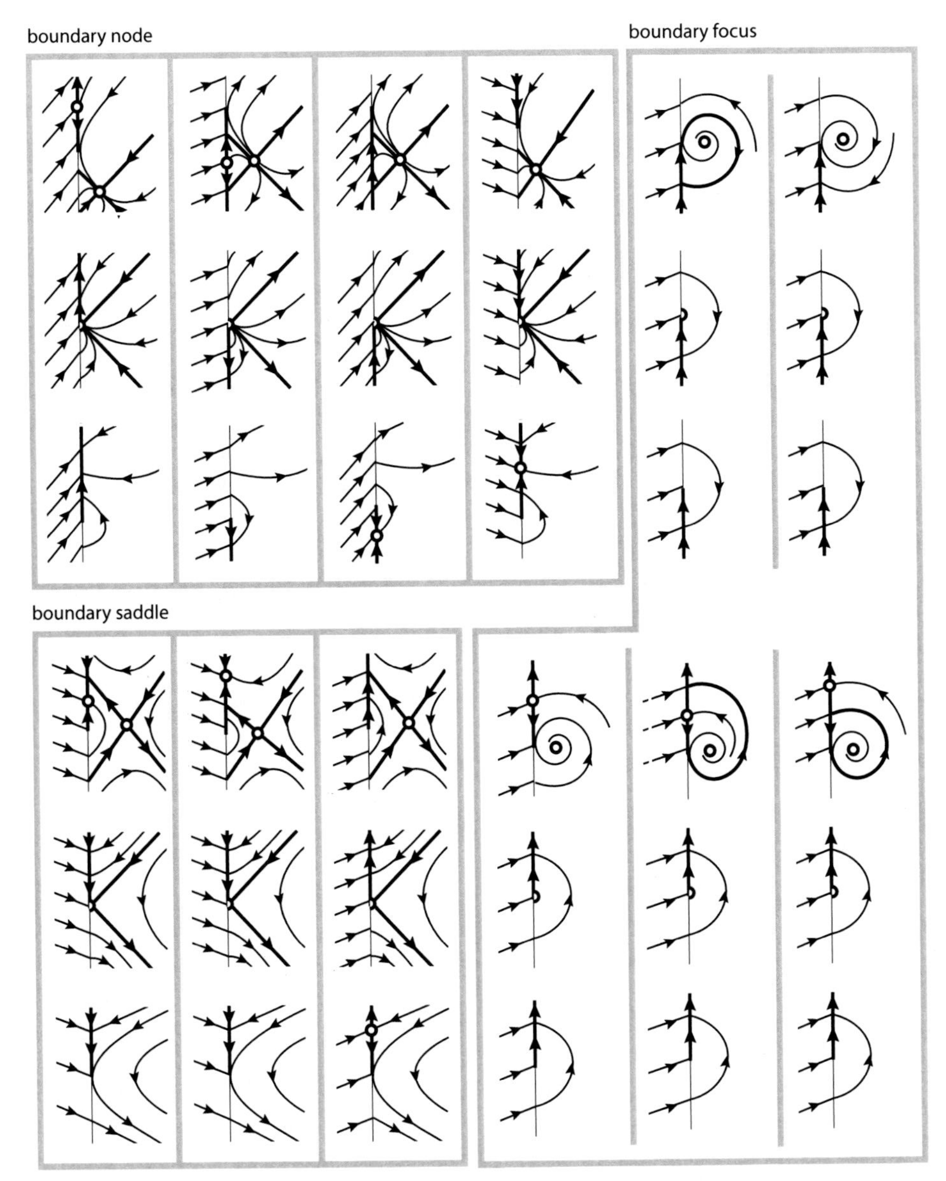

Figure 1.31: Codimension 1 boundary equilibria and their unfoldings.

Each triplet shows the unfolding as μ changes sign. The middle portrait in each shows the bifurcation point – the boundary equilibrium. You can find most of these (some were missed!) in [67] and later references. You'll find the bifurcation points themselves (all of them!) only in Filippov's book [29].

These unfold as a single parameter (μ) changes, so they are codimension one bifurcations.

Filippov [29] also classified the bifurcation points of boundary equilibria, of tangencies, and of more exotic things like line singularities. Little headway has been made so far, however, in extending such bifurcation studies into higher dimensions.

1.11.3 Global bifurcations and tangencies

Global bifurcations consist of orbital connections between singularities, i.e., between equilibria, sliding equilibria, and tangencies. The latter of course come into play only in piecewise-smooth systems, and give rise to global discontinuity-induced bifurcations.

These may arise when distinguished orbits (those already connected to an equilibrium or periodic orbit, for example) have connections to visible or invisible tangencies. In fig. 1.32, the ★ denotes connection some general singularity, perhaps another equilibrium, sliding equilibrium, or tangency. The connection need not be simple, however, and could consist of orbit segments that cross or slide on other switching surfaces.

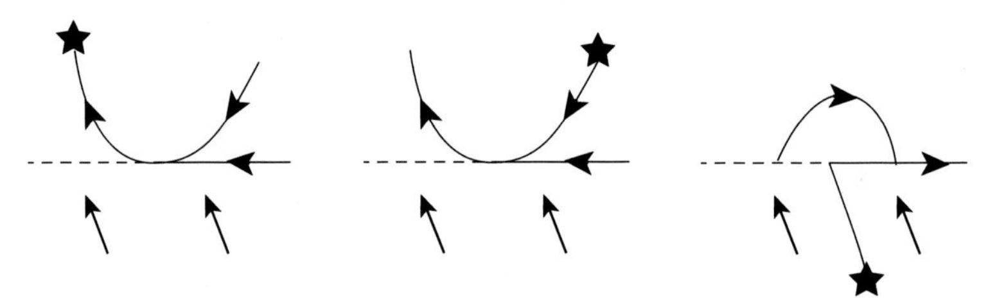

Figure 1.32: Connections to a visible (left) or invisible (right) tangency are structurally unstable.

Figure 1.33 shows various connections that give rise to bifurcations, though many more are possible, particularly when we consider multiple dimensions, and multiple switches.

Cases (i)–(iii) of fig. 1.33 show heteroclinic connections between an equilibrium outside the switching surface and a tangency, (iv) shows heteroclinic connection between an equilibrium outside the surface and a sliding equilibrium. The last two cases (v)–(vi) show two ways that a visible tangency can facilitate a homoclinic connection to an equilibrium or a sliding equilibrium.

Under perturbation of any of these the connection typically will be broken, and a bifurcation takes place. See [21, 67] for further examples.

A visible tangency may connect to itself, forming a periodic orbit. Let us unfold this particular bifurcation, known as a *grazing-sliding* bifurcation, shown in fig. 1.34. On one side of the bifurcation is a smooth limit cycle, on the other side is a so-called "stick-slip" oscillation.

When repelling sliding is involved, bifurcations can take another, more dramatic form, shown in fig. 1.35. We call these sliding *explosions*.

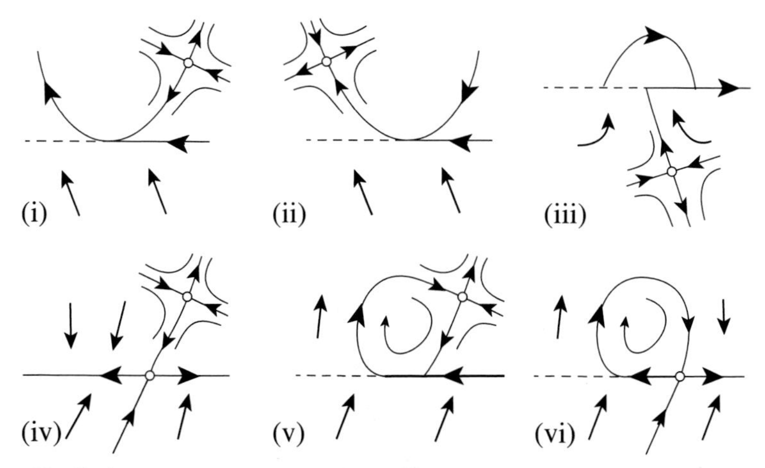

Figure 1.33: Codimension one connections: Heteroclinic connection between a saddle and a visible tangency (i)–(ii), a saddle and invisible tangency (iii), a saddle and sliding saddle (iv). Homoclinic connection via visible tangency to a saddle (vi) or a sliding saddle (vii).

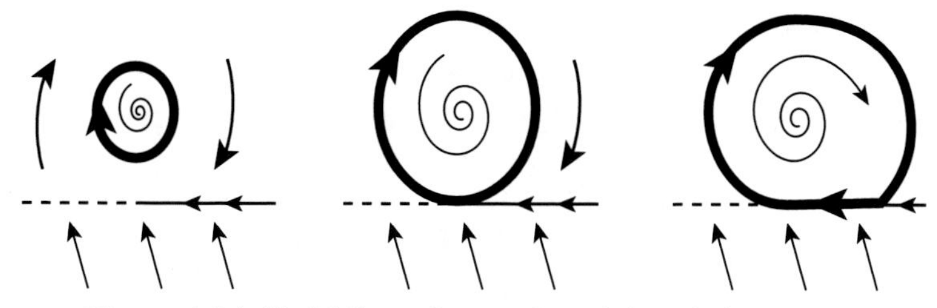

Figure 1.34: Unfolding of a grazing-sliding bifurcation.

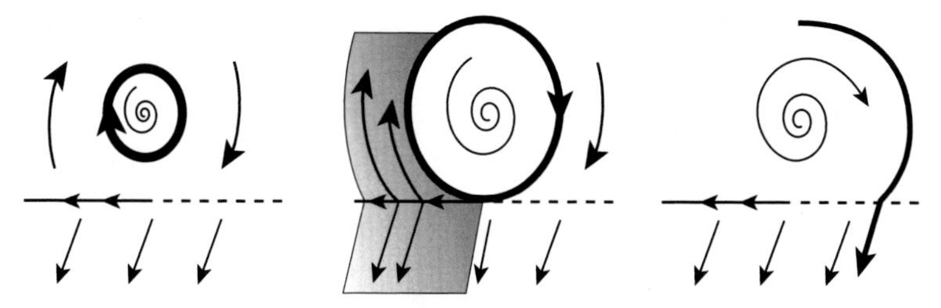

Figure 1.35: Unfolding of a grazing-sliding explosion [58].

1.12 Determinacy-breaking

An explosion is one example of *determinacy-breaking*, a general phenomenon that occurs when the flow is somehow able to enter a region of repelling sliding.

The only robust way this can happen is when a sliding region changes attractivity, so the flow passes from attracting sliding to repelling sliding. (Notice in the grazing sliding explosion above that the explosion happens only at a special parameter value, i.e., it doesn't happen at a general parameter value, hence it is not 'robust'.)

One way this can happen is at a two-fold singularity (see earlier example and [58]).

A simpler way is at a switching surface made of two intersecting manifolds. For example the scenario shown in fig. 1.36.

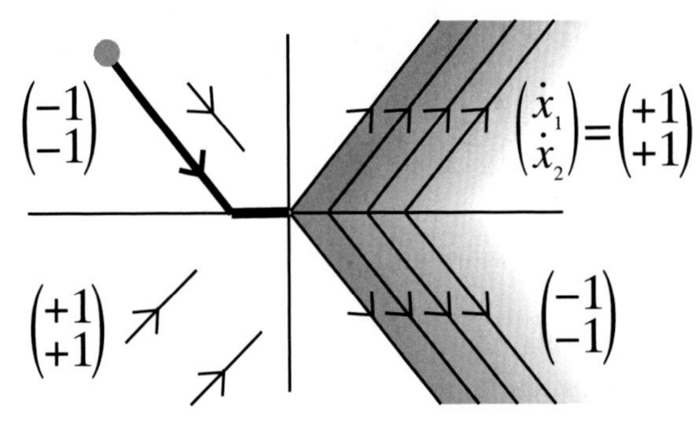

Figure 1.36: Determinacy-breaking at a switching intersection. The system has two switches creating four regions. The loss of determinacy at the intersection can be partially resolved using layer analysis, see [61].

Because of the presence of determinacy-breaking points like these, piecewise-smooth systems are *almost* deterministic, at best. The techniques of combinations and switching layers that we have explored here resolve the non-uniqueness of discontinuous systems *as far as possible* and no further, leaving determinacy-breaking points as an essentially new fundamental phenomenon in dynamical systems theory.

1.13 Hidden attractors

A hidden attractor is one that exists inside the switching surface, but whose existence or behaviour is determined by nonlinear (or multi-linear) dependence on switching multipliers, and therefore is revealed only by switching layer analysis.

This can lead to strange seeming behaviour in systems that, from outside the switching surface, appear simple.

Consider

$$\begin{aligned}
\dot{x}_1 &= 5(\lambda_2 - \lambda_1) - 75x_1, \\
\dot{x}_2 &= -\lambda_1 - 15\lambda_1\lambda_3 - \tfrac{1}{2}\lambda_2 - 75x_2, \\
\dot{x}_3 &= 15\lambda_1\lambda_2 - \tfrac{4}{3} - \tfrac{4}{3}\lambda_3 - 75x_3, \\
\dot{x}_4 &= sw_1 - 75x_4,
\end{aligned} \tag{1.102}$$

where $\lambda_j = \text{sign}(x_j)$ for $j = 1, 2, 3$.

This has three switches, and eight regions, but the dynamics is seemingly almost trivial. Outside the switching surface each row looks like $\dot{x}_i = \text{const} - 75x_i$, the value of the 'const' term just jumps across the switching surfaces, giving strong attraction towards $\frac{1}{75} \times \text{const}$, which ultimately results in collapse towards the origin. Once at $(x_1, x_2, x_3) = (0, 0, 0)$, where all three switches intersect, however, things become interesting.

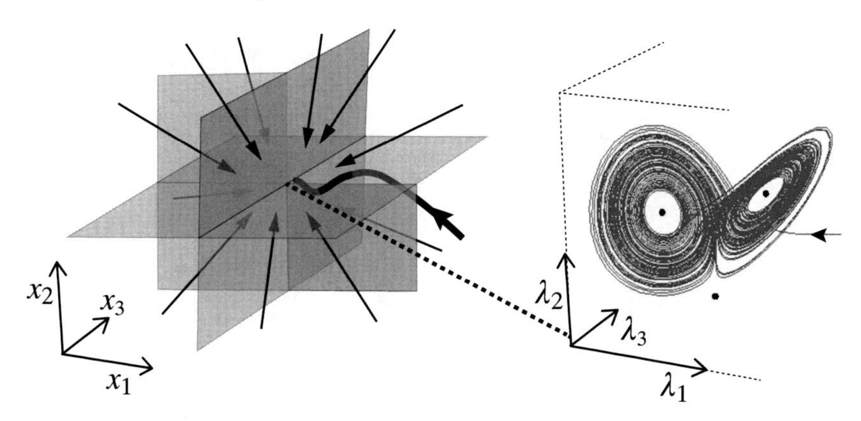

Figure 1.37: Hidden Lorenz attractor at the intersection of three switches. This example was inspired by gene regulatory networks [72].

When $(x_1, x_2, x_3) = (0, 0, 0)$, the layer system, letting in this example $\varepsilon_i = \varepsilon$ for $i = 1, 2, 3$, becomes the Lorenz system in $(\lambda_1, \lambda_2, \lambda_3)$,

$$\begin{aligned} \varepsilon\dot{\lambda}_1 &= 5(\lambda_2 - \lambda_1), \\ \varepsilon\dot{\lambda}_2 &= -\lambda_1 - 15\lambda_1\lambda_3 - \tfrac{1}{2}\lambda_2, \\ \varepsilon\dot{\lambda}_3 &= 15\lambda_1\lambda_2 - \tfrac{4}{3} - \tfrac{4}{3}\lambda_3, \end{aligned} \tag{1.103}$$

so the three switching multipliers behave chaotically inside the switching layer $(\lambda_1, \lambda_2, \lambda_3) \in (-1, +1)^3$. The variables x_1, x_2, x_3, remain at zero. The chaotic dynamics does affect the global system, however. Since the variable x_4 is coupled to the λ_1 switching multiplier, it will follow λ_1's chaotic trajectory.

1.14 Hidden bifurcations

Hidden attractors can undergo their own bifurcations, including any bifurcations that are possible in smooth systems, and many more that remain to be discovered. Consider

$$\begin{aligned} \dot{x}_1 &= \tfrac{1}{2}(1 - \lambda_1\lambda_2) - \nu_1(x_1 + \theta_1) \\ \dot{x}_2, &= \frac{1}{4}(3 - \lambda_1 - \lambda_2 - \lambda_1\lambda_2) - \nu_2(x_2 + \theta_2), \end{aligned} \tag{1.104}$$

where $\lambda_j = \text{sign}(x_j)$ for $j = 1, 2$, shown in fig. 1.38.

This has two switches that intersect at the origin $(0,0)$. There is attraction towards $(0,0)$ from some directions, either directly or via sliding, but also repulsion from $(0,0)$ via sliding. To find out what the flow does we must look inside the switching layer, shown on the right of fig. 1.38.

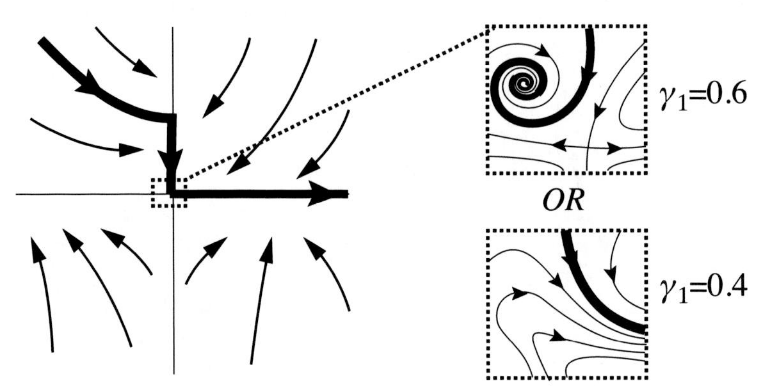

Figure 1.38: Hidden saddle-node bifurcation at the intersection of two switches. This example was derived from models of gene regulatory networks [26].

The layer system for $x_1 = x_2 = 0$ undergoes a saddle-node bifurcation as ν_1 changes value. There is an attracting focus for higher values of ν_1, which traps solutions at $(0,0)$, but as ν_1 decreases the focus collides with a saddle and annihilates, leaving no attractor, so solutions pass through $(0,0)$ onto the right-hand half line of the switching surface.

1.15 Moving forward

Piecewise-smooth dynamics has a long history, but it is still a young, rapidly growing and exciting area full of new ideas and open challenges. Let me just end by picking out – quite arbitrarily – a few of the many highlights of recent years.

Two big breakthroughs from recent years:

- The two-fold singularity – Is it stable or not? Is it an attractor or not? For the beginnings of the problem see [29, 92] and the CRM Intensive Research Program on Complex Nonsmooth System 2007. This was eventually solved around 2008–2011, extended to many dimensions in 2013, and to many switches in 2015. See [64] for a summary.
- Naive ideas of extending Hilbert's 16th problem to nonsmooth systems were thrown wide open when it was shown in 2015 that infinitely many cycles are possible even in a piecewise linear system [71].

Two big challenges outstanding:

- Regularization / Non-ideal switching – how do we deal with effects of smoothing, hysteresis, stochastics, discretization, delay, and others we perhaps have not yet considered. What do discontinuities in real systems, throughout physics, chemistry, engineering and the life sciences, look like as we delve deeper into their modeling?
- Higher dimensions — what new attractors/bifurcations/chaos or entirely new phenomena appear in higher dimensions? What new concepts do we need to describe them?

The major applications of piecewise-smooth dynamics now include modeling for:

- engineering and environment — climate, power control, economy, process engineering, earthquakes, biscuits, robotics, classical mechanics, chemical reactions, superconductors, friction/impact, etc. . .
- life sciences — cell biology, social behaviour, neuroscience, ecology, genetic regulation, demography, etc. . .

There are many more breakthroughs, open challenges, and emerging applications that could be listed here, but I will leave them for you to discover for yourselves. When given at the CRM, this course ended with a discussion of *why* so many systems involve discontinuities, why they are more than a crude modeling tool and are in fact a subtle asymptotic phenomenon, and why so many systems are well described by the piecewise-smooth model of switching. This discussion has been summarized in the CRM Extended Abstract Series article titled *Why nonsmooth?* [63].

Chapter 2

Piecewise-smooth Maps

2.1 Introduction to maps

This course is about piecewise-smooth maps. If the phase space (typically $\mathbb{R}^n$) is partitioned into N disjoint open regions such that the union of the closures of these regions is the whole space, then a piecewise-smooth map is a map on this partition which is defined by a different smooth function on each region. Note that a piecewise-smooth map may be discontinuous across boundaries, or it may be continuous but the Jacobian matrix is discontinuous. Other classes exist, but these two form the basis for most studies. The decision about how to define dynamics on the boundaries of the regions can be a bit awkward and will involve us in some little technical issues later.

Given this description you may think that these maps are really rather special and uninteresting, so the first question you should ask about the study of piecewise-smooth maps is: why bother?

2.1.1 Piecewise-smooth maps are interesting

Piecewise-smooth maps are interesting. So interesting that the ideas, examples and techniques involved in their study have been rediscovered by different groups at different times. This is in some sense irritating (it is hard to know what has been done, and the same phenomenon is called by a different name in different groups making comparisons hard), but it also emphasises how central piecewise-smooth systems are in the study of dynamics. Despite the range of modern applications the area is still seen from the outside as quite a narrow interest group. On the other hand, as Mike Field says, engineers (both mechanical and electronic) have spent the last 50 years working with systems containing jumps, whilst the dynamical systems community has spent the last 50 years perfecting the theory of smooth dynamical systems. It is time for a change!

P. Glendinning, M. R. Jeffrey, *An Introduction to Piecewise Smooth Dynamics*, Advanced Courses in Mathematics - CRM Barcelona, https://doi.org/10.1007/978-3-030-23689-2_2

The list below gives an idea of the groups that have been interested in piecewise-smooth systems. It is neither complete, nor accurate (and I apologise in advance to those who think I have put them in the wrong group), but it gives an impression of the diversity of approaches and interests in the area.

- (Mechanics, from 1990s) Budd, di Bernardo, Champneys, Dankowitz, Nordmark, Hogan (from 1980s!).
- (Electronics, applied dynamical systems, 1990s) Banerjee, Grebogi, Nusse, Ott, Yorke.
- (Ergodic Theory, from 1980s) Young, Misieurewicz, Chernov, Pesin, Jakobson, Newhouse; Buzzi, Keller, Saussol, Tsujii.
- (Classification of flows on manifolds and rational billiards, from 1960s) Viana and the interval exchange map community.
- (Non-invertible maps, from 1980s) Avrutin, Gardini, Lozi, Mira, Schanz, Shushko.
- (Homoclinic bifurcations, 1970s) Gambaudo, Glendinning, Holmes, Lorenz, Procaccia, Tresser.
- (Structure Theorems, 1970s) Alseda, Guckenheimer, Llibre, Milnor, Misiurewicz, Rand, Thurston, Williams.
- (Rotations, 1980s) Herman, Kadanoff, Keener, Lanford, Rhodes, Thompson.
- (Modern Nonsmooth, 2000s) Colombo, Granados, Jeffrey, Simpson.

I could go on, but you get the point.

2.1.2 Motivating examples

There are a number of standard examples that give a sense of the many models that can be described via piecewise-smooth systems. Here are a few.

A bouncing ball

Suppose a ball is dropped and starts bouncing. Let v_n be the speed (upwards) immediately after the nth bounce at time t_n. It will rise to height h_n with $2gh_n = v_n^2$ after time v_n/g and then return to the ground at time $t_{n+1} = t_n + 2v_n/g$ with the same speed v_n. The collision with the ground instantaneously reverses the direction of the the velocity and reduces its magnitude by a factor $r \in (0,1)$, so $v_{n+1} = rv_n$.

This impacting system therefore has a jump in the velocity at each collision but the equations describing the change over each bounce are

$$v_{n+1} = rv_n, \quad t_{n+1} = t_n + 2v_n/g, \quad 0 < r < 1.$$

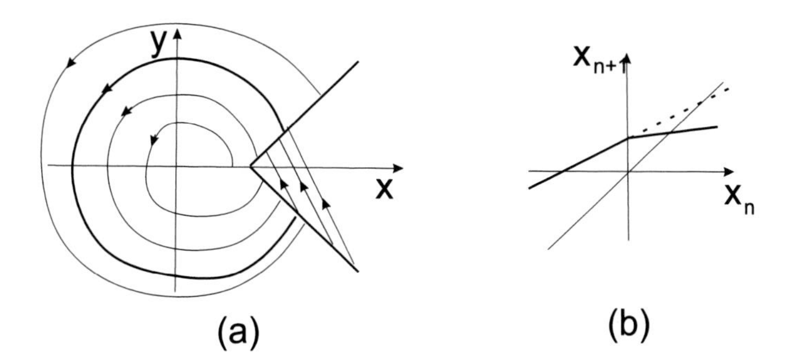

Figure 2.1: Corner collision bifurcations. (a) Phase portrait just after the bifurcation; (b) return map on the positive x-axis showing a change in slope at $x = 1$. The dotted line indicates the extension of the map (2.5) into $x > 1$.

Thus although the dynamics is piecewise-smooth with jumps in phase space, the modelling map is smooth and is *not* of the sort that will concern us here. Indeed, they can be solved:

$$
\begin{aligned}
v_n &= r^n v_0, \\
t_n &= t_0 + \tfrac{2v_0}{g}(1 + r + r^2 + \cdots + r^{n-1}) = t_0 + \tfrac{2v_0(1-r^n}{g(1-r)}.
\end{aligned}
\tag{2.1}
$$

As $n \to \infty$, $t_n - t_0 \to \frac{2v_0}{g(1-r)}$, i.e., there are an infinite number of bounces in finite time for this model system.

A corner collision bifurcation

A corner collision bifurcation occurs in a two-dimensional piecewise-smooth flow if one of the regions on which the smooth flows are defined takes the form of a wedge. The general case is described in [20] and here we consider an example similar to the example in [21].

Let W be the wedge-shaped region

$$W = \{(x, y) \mid -x + 1 < y < x - 1,\ x > 1\}, \tag{2.2}$$

and suppose that the dynamics in W is defined by the constant differential equation

$$\dot{x} = -a, \quad \dot{y} = 1, \quad 0 < a < 1. \tag{2.3}$$

Since solutions lie on straight lines with gradient $-\frac{1}{a}$, the condition $0 < a < 1$ implies that every solution starting at $(x_0, -x_0 + 1)$, $x_0 > 1$, on the lower boundary of W strikes the upper boundary in finite time at with an x-coordinate between 1 and x_0, see fig. 2.1a.

Now suppose that outside W (strictly speaking, outside the closure of W) the flow is determined by the Hopf bifurcation normal form, with, in polar coordinates (r,θ)

$$\dot{r} = r(\mu - r^2), \quad \dot{\theta} = \omega, \tag{2.4}$$

with $\omega > 0$ and μ a real parameter. The Hopf bifurcation occurs as μ passes through zero, creating a stable periodic orbit of radius $\sqrt{\mu}$ in $\mu > 0$. Provided $\mu < 1$ this orbit does not intersect W. Since it is stable, the local behaviour is determined by the Floquet multiplier A, $0 < A < 1$, of the orbit (see, e.g., [42]) and a return map on the positive x-axis for solutions that do not intersect W is, ignoring quadratic terms and higher,

$$x_{n+1} = \sqrt{\mu} + A(x_n - \sqrt{\mu}). \tag{2.5}$$

Now suppose that μ is a bit larger than 1. Then the return map (2.5) remains valid if $x_n < 1$, but if $x_n > 1$ then the image is deflected a little to the left (i.e., to smaller values of x, and so $x_{n+1} = \sqrt{\mu} + B(x_n - \sqrt{\mu})$ for some B with $0 < B < A$.

Hence the modified return map has a kink as shown in fig. 2.1b and the map takes the approximate form

$$x_{n+1} = \begin{cases} \sqrt{\mu} + A(x_n - \sqrt{\mu}) & \text{if } x_n < 1, \\ \sqrt{\mu} + B(x_n - \sqrt{\mu}) & \text{if } x_n > 1. \end{cases} \tag{2.6}$$

Since $0 < B < A < 1$ the periodic orbit continues to exist as μ increases through unity, but its Floquet multiplier changes. The dynamics of monotonic maps such as (2.6) are described in Section 2.2.1, Lemma 2.2.2. If the signs of A and B are different, and if they can have modulus greater than on, then much greater complexity is possible, and this is described in greater detail in [56, 57].

The Lorenz semiflow

The Lorenz equations provide one of the early examples of differential equations with chaotic attractors (although the proof that the attractor really is chaotic is relatively recent). Guckenheimer and Williams [54, 100] developed mathematical abstractions of the equations, assuming that the flow lies on the branched manifold of fig. 2.2a. In this case the chaos is due to solutions falling on one or other side of the stable manifold of a saddle and being swept round a loop to the left or to the right. The return map of the model flow takes the form shown in fig. 2.2b. It has a discontinuity at the origin (the stable manifold of the saddle) and the slope goes to infinity like $|x|^\alpha$, $0 < \alpha < 1$ at the point of discontinuity. We will return to maps like these in later sections.

2.1.3 Phenomenology

In many cases the interest is not in a particular map, but in a family of maps. Thus many results aim to describe the structure of dynamics as a function of

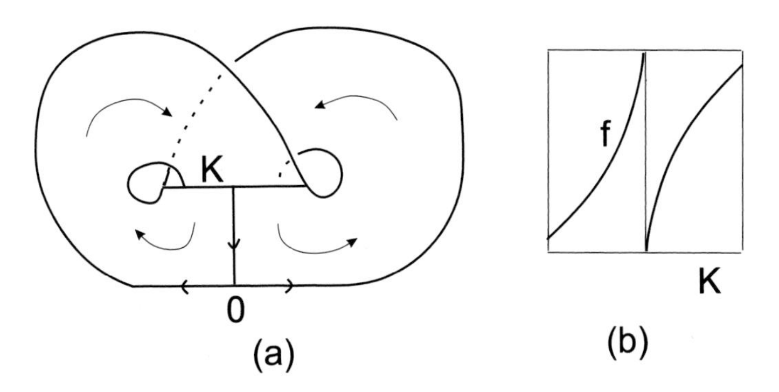

Figure 2.2: Lorenz semiflow and the associated one-dimensional map, after [100]. (a) The branched manifold on which the semiflow is defined; (b) the associated return map on the interval K on which the two branches are glued together.

parameter, i.e., the bifurcation theory of these maps. One feature that stands out in piecewise-smooth systems because it is not present in smooth systems is period-adding. In period-adding bifurcations there is a sequence of bifurcations in which a constant is added to the period of the orbit at each bifurcation. Sometimes the bifurcations are clean, in the sense that there are no intermediary bifurcations, and sometimes more complicated, with bands of chaos separating the added orbits. Nordmark's square root map provides one such example ([78] and Section 2.2.3). The Nordmark (or square root) map is continuous everywhere and differentiable except at a single point:

$$x_{n+1} = \begin{cases} \mu + ax_n & \text{if } x < 0, \\ \mu - b\sqrt{x_n} & \text{if } x \geq 0. \end{cases} \tag{2.7}$$

This map is analyzed in Section 2.2.3, and another way in which period adding sequences can be generated is discussed in Section 2.5b.

In many circumstances more than one parameter is present, and the sensitivity to changes in the parameter can be mind-boggling. The results of careful numerical simulations such as those by Avrutin et al [8] show very complicated regions of dynamics in examples involving two parameters. This level of complexity makes it hard to decide what feature is worth concentrating upon in any analysis.

These observations bring out an important feature of the non-smooth world. The number of possible behaviours seems to be huge, and the complexity of the bifurcation diagrams and their sensitivity to changes in other parameters can be quite bewildering. For the mathematician used to tidy classifications this can be a problem. One of the recurring themes of this lecture series is that 'less is more'. In a world of extraordinary complexity it may not be either useful or possible to obtain a complete list of theoretical possibilities, and that a less complete description may be more useful.

2.1.4 Less is more

The comments of Section 2.1.3 and the results of, e.g., [51] suggest that the level of complexity of bifurcations in even quite simple piecewise-smooth systems is much greater than that for smooth flows. In the theory of smooth systems it is standard to give quite general bifurcation theorems which reflect the important local features of the dynamics. It seems likely that there is a proliferation of cases for piecewise-smooth systems which means that detailed bifurcation theorems are much less useful, and it is then a matter of judgement about how much detail should be given.

These lectures reflect this attitude. I will use some standard examples to illustrate techniques rather than attempt to provide a detailed description of every bifurcation in the literature. This might make the use of a small number of examples appear unbalanced, but (I hope) that the techniques described here can be applied to many of the examples that might be met in applications. For more detail of 'less is more' see [49, 50].

2.2 Smooth theory

By definition a piecewise-smooth system is smooth in regions, so any dynamics that does not interact with a boundary can be described using smooth theory. This includes the existence and stability of fixed points and periodic orbits in smooth regions and their bifurcations (in Section 2.3.5 we will look at some elementary new bifurcations involving the boundary).

A smooth map is simply a smooth function $f\colon \mathbb{R}^n \to \mathbb{R}^n$ which generates dynamics via the difference equation

$$x_{n+1} = f(x_n). \tag{2.8}$$

Thus given a point x the *orbit* of x is generated by applying f, creating the sequence

$$(x, f(x), f^2(x), f^3(x), \dots)$$

where $f^n(x) = f(f^{n-1}(x))$, $n \geq 2$. Thus whenever we write f^n we mean the nth iterate of f,

$$f^n = f \circ f \circ \dots \circ f \quad (n \text{ times})$$

and *not* the nth power of $f(x)$ which will be denoted by $[f(x)]^n$.

One of the central ideas in dynamical systems is that of invariance.

Definition 2.2.1. Given a map $f\colon \mathbb{R}^n \to \mathbb{R}^n$, a set $S \in \mathbb{R}^n$ is invariant if $f(x) \in S$ for all $x \in S$ (often written as $f(S) \subseteq S$).

This definition is sometimes called forward invariance, and if f is a homeomorphism then a set S is both forward and backward invariant (often abbreviated to invariant if the context is clear) if $f(S) = S$, so $x \in S$ implies $f^{-1}(x) \in S$ and $xf(x) \in S$.

The simplest (geometrically) invariant sets are fixed points. A fixed point of a smooth map $f: \mathbb{R}^n \to \mathbb{R}^n$ is a solution of

$$x = f(x) \tag{2.9}$$

and it is stable (or more accurately, linearly stable) if all the eigenvalues of the Jacobian matrix

$$Df(x) = \begin{pmatrix} \frac{\partial f_1}{\partial x_1} & \frac{\partial f_1}{\partial x_2} & \cdots & \cdots & \frac{\partial f_1}{\partial x_n} \\ \frac{\partial f_2}{\partial x_1} & \frac{\partial f_2}{\partial x_2} & \cdots & \cdots & \frac{\partial f_2}{\partial x_n} \\ \vdots & \vdots & \vdots & \vdots & \vdots \\ \frac{\partial f_n}{\partial x_1} & \frac{\partial f_n}{\partial x_2} & \cdots & \cdots & \frac{\partial f_n}{\partial x_n} \end{pmatrix} \tag{2.10}$$

evaluated at the fixed point lie inside the unit circle (i.e., have modulus less than 1).

A point is periodic of period p if

$$x = f^p(x) \tag{2.11}$$

where $f^p(x) = f(f^{p-1}(x))$, i.e., it denotes the pth iterate of f,

$$f^p = f \circ f \circ \cdots \circ f \quad (p \text{ times}),$$

and not the pth power of $f(x)$ which we will denote by $[f(x)]^p$ or similar. If x is a point of period p then the periodic orbit containing x is

$$\{x, f(x), \ldots, f^{p-1}(x)\}$$

and if all the points are distinct then it is sometimes worth emphasising that p is the minimal possible period of the orbit (though usually this is left unstated). Note that if x has period p, then it also has period mp for all $m > 1$.

Since a periodic point can be viewed as a fixed point of f^p, the linear stability of a periodic orbit is determined by the eigenvalues of the Jacobian matrix

$$Df^p(x) = Df(f^{p-1}(x))Df(f^{p-2}(x))\cdots Df(x).$$

Bifurcations occur is an eigenvalue passes through the unit circle, so there are three generic cases: a simple eigenvalue of +1, a simple eigenvalue of +1, or a pair of simple eigenvalues $e^{\pm i\theta}$, $\theta \neq m\pi$, $m \in \mathbb{Z}$.

The Centre Manifold Theorem implies that these cases can be classified in the same way regardless of the dimension of the phase space, a feature that is not true of piecewise-smooth bifurcations). An eigenvalue of +1 implies that for small changes of parameter there is typically a saddle-node bifurcation in which as a parameter is varied a pair of fixed points come together at the bifurcation parameter and do not exist thereafter. Symmetries or the non-generic vanishing of some derivatives of the Taylor expansion of the map can imply that a saddlenode bifurcation does not happen and there may be a transcritical bifurcation (exchange of stability) or a pitchfork bifurcation.

An eigenvalue of -1 leads to a period-doubling bifurcation: as parameters vary a fixed point changes stability at the bifurcation value and an orbit of period two is created. If this period two orbit is stable it is called a supercritical period-doubling bifurcation.

A pair of eigenvalues $e^{\pm i\theta}$, $\theta \neq m\pi$, $m \in \mathbb{Z}$ leads to a Hopf, or Niemark–Sacker bifurcation. The fixed point changes stability and an invariant curve bifurcates on which there can be other attractors (e.g., periodic orbits) near resonances when θ is a rational multiple of 2π.

2.2.1 Markov partitions and chaos

There are a number of results which make the analysis of one-dimensional systems significantly easier than higher-dimensional dynamics. The first result describes the dynamics of monotonic maps.

Lemma 2.2.2. *Suppose $f: \mathbb{R} \to \mathbb{R}$ is a continuous map. If f is increasing then every bounded orbit is either a fixed point or tends to a fixed point. If f is decreasing, then every bounded orbit is either a fixed point or a point of period two or tends to a fixed point or a point of period two.*

Proof. Suppose that f is increasing, i.e., $x < y$ implies that $f(x) \leq f(y)$. Take $x \in \mathbb{R}$. Then either

$$f(x) = x, \quad \text{or} \quad f(x) > x, \quad \text{or} \quad f(x) < x.$$

In the first case x is a fixed point. In the second case $x < f(x)$ implies that $f(x) \leq f^2(x)$ using the increasing property, and hence by induction $(f^k(x))$ is an increasing sequence. It is therefore either unbounded or bounded above. If it is bounded above, then the sequence tends to a limit, ℓ, and hence $(f^{k+1}(x))$ tends to $f(\ell)$. But the two sequences are the same (by continuity of f) and hence $\ell = f(\ell)$, i.e., ℓ is a fixed point. In the third case $f(x) < x$ implies that $f^2(x) \leq f(x)$ and so $(f^k(x))$ is a decreasing sequence. It is therefore unbounded or bounded below, in which case by the same argument as in the second case the limit is a fixed point.

If f is decreasing, then $x < y$ implies that $f(x) \geq f(y)$ and hence $f^2(x) \leq f^2(y)$. Hence f^2 is increasing and since fixed points of f^2 are either fixed points or points of period two for f the second part of the lemma holds. □

Lemma 2.2.2 describes simple behaviour — we now describe how to treat some chaotic dynamics. The first idea is the transition matrix. Throughout this section, f will be a continuous map $f: \mathbb{R} \to \mathbb{R}$.

Definition 2.2.3. If J and K are is a closed intervals, then J f-covers K if $K \subseteq f(J)$.

Lemma 2.2.4. *If J f-covers itself, then J contains a fixed point of f.*

Proof. Let $J = [a,b]$. Since J f-covers itself there exist y and z in $[a,b]$ such that $f(y) \le a$ and $f(z) \ge b$. Let $g(x) = f(x) - x$ which is also continuous and $g(y) \le 0$ and $g(z) \ge 0$. Applying the Intermediate Value Theorem to g on the interval between x and y there exists u such that $g(u) = 0$, i.e., u is a fixed point of f. □

Definition 2.2.5. Let $J_1, \dots, J_m$ be closed intervals with disjoint interiors. A Markov graph of f is a directed graph with vertices $1, \dots, m$ and a directed edge from i to j iff J_i f-covers J_j. The transition matrix associated with this graph is the $m \times m$ matrix T with

$$T_{ij} = \begin{cases} 1 & \text{if } J_i \ f\text{-covers } J_j, \\ 0 & \text{otherwise.} \end{cases}$$

A path in a directed graph is an ordered sequence of vertices $a_0 a_1 \cdots a_k$ such that there is a directed edge from a_i to a_{i+1} for each $i = 0, \dots, k-1$. The length of the path is the number of edges traversed (i.e., k in the example). Note that if there is a path from a_0 to a_k of length k if and only if $T^k_{a_0 a_k} \neq 0$.

Lemma 2.2.6. *If there is a path of length k from $a_0 \cdots a_k$ in the Markov graph then there exists a closed interval $L \subset J_{a_0}$ such that $f^k(L) = J_{a_k}$ and $f^r(L) \subseteq J_{a_r}$.*

Proof. The proof is by induction on k.

If $f(J_{a_0}) \subseteq J_{a_1}$, then since f is continuous there exists $L \subseteq J_{a_0}$ such that $f(L) = J_{a_1}$. (If this is not obvious, look at the interior of J_{a_1} and note f^{-1} of an open interval is a union of open intervals.)

Now suppose that the lemma is true for $k = m$ and consider a path of length $m+1$, $a_0 \cdots a_{m+1}$. By the induction hypothesis, since $a_1 \cdots a_{m+1}$ is a path of length m there exists $L' \subseteq J_{a_1}$ such that $f^m(L') \subseteq J_{a_{m+1}}$ and $f^r(L') \subseteq J_{a_{r+1}}$, $r = 1, \dots, m$.

Since J_{a_0} f-covers J_{a_1}, J_{a_0} f-covers L' and hence there exists $L \subseteq J_{a_0}$ such that $f(L) = L'$. A quick check confirms that $f^{k+1}(L) = f^k(L')$ and so L has the desired property. □

Corollary 2.2.7. *If $a_0 a_1 \cdots a_p$ is a path of length p with $a_0 = a_p$, then f has a periodic orbit of period p.*

Corollary 2.2.8. *If $a_0 a_1 \cdots$ is an infinite path in the Markov graph, then there exists $x \in J_{a_0}$ such that $f^r(x) \in J_{a_r}$ for all $r \ge 0$.*

Proof. Take an infinite intersection of nested closed intervals L of Lemma 2.2.6 for each finite path $a_0 \cdots a_r$. □

This is the basic tool for proving classic theorems such as Sharkovskii's Theorem. It also provides a motivation for the definition of a one-dimensional horseshoe.

Definition 2.2.9. f has a horseshoe if there exist closed intervals J_0 and J_1 with disjoint interiors such that J_0 f-covers both J_0 and J_1 and J_1 f-covers both J_0 and J_1.

Theorem 2.2.10. *If f has a horseshoe, then for any sequence of 0s and 1s, $a_0 a_1 \cdots$, there exists $x \in J_{a_0}$ such that $f^r(x) \in J_{a_r}$ for all $r > 0$.*

This is sometimes described as f having dynamics equivalent to a full shift on two symbols.

Note that these results only need f to be continuous on the intervals J_k; what happens between these intervals is immaterial. This means that the methods are often applicable in piecewise-smooth systems.

Definition 2.2.11. A continuous map of the interval f is chaotic if there exists $n \geq 1$ such that f^n has a horseshoe.

An Example

An example of how this theory can be applied is shown in fig. 2.3a. The map $F: [0,1] \to [0,1]$ from fig. 2.3a is continuous and PWS via smooth maps defined on each of the regions of the partition $0 < x_1 < x_2 < 1$. The map is monotonic on each interval defined by the partition and

$$F(0) = F(x_2) = x_2, \quad F(x_1) = 1, \quad \text{and} \quad F(1) = 0.$$

Let $I_1 = [0, x_1]$, $I_2 = [x - 1, x_2]$, and $I_3 = [x_2, 1]$. Then $F(I_1) = F(I_2) = I_3$ and $F(I_3) = I_1 \cup I_2$ and so the directed graph associated with the map is as shown in fig. 2.3b and the associated transition matrix is

$$T = \begin{pmatrix} 0 & 0 & 1 \\ 0 & 0 & 1 \\ 1 & 1 & 0 \end{pmatrix} \quad \text{with} \quad T^2 = \begin{pmatrix} 1 & 1 & 0 \\ 1 & 1 & 0 \\ 0 & 0 & 1 \end{pmatrix}. \tag{2.12}$$

Thus $F^2(I_k) = I_1 \cup I_2$, $k = 1, 2$ and so F is chaotic. In fact, any sequence made up of concatenations of 13 and 23 (and shifts of such sequences) is allowed by the graph.

2.2.2 Continuous maps of the interval

A piecewise-smooth map without discontinuities is a continuous map of the interval and hence any general result for continuous maps holds for piecewise-smooth maps without discontinuities. There is one remarkable result for such maps that is worth recalling (we will not give the proof, though it only uses the ideas of Markov partitions from Section 2.2.1). Sharkovskii's Theorem describes how the existence of a periodic orbit of a given period can imply the existence of periodic orbits of other periods.

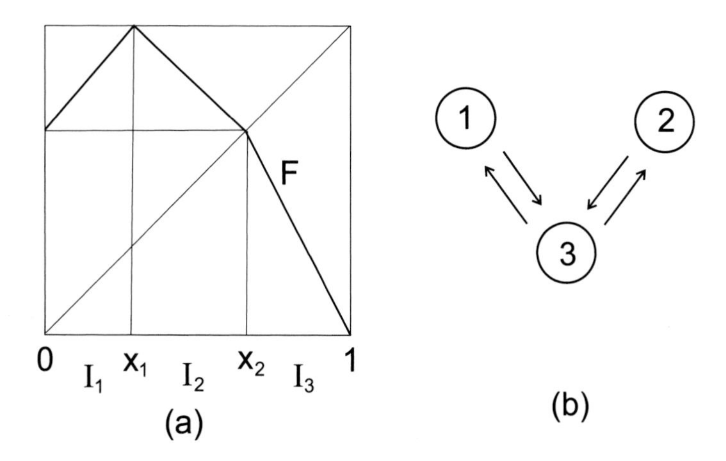

Figure 2.3: (a) The continuous PWS map F; and (b) the associated directed graph.

Consider the complete order "$\prec$" (the *Sharkovskii order*) on the positive integers defined by

$$\begin{aligned}
&1 \prec 2 \prec 4 \prec \cdots \prec 2^n \prec 2^{n+1} \prec \cdots \\
&\cdots 2^{n+1}.11 \prec 2^{n+1}.9 \prec 2^{n+1}.7 \prec 2^{n+1}.5 \prec 2^{n+1}.3 \prec \cdots \\
&\cdots 2^n.11 \prec 2^n.9 \prec 2^n.7 \prec 2^n.5 \prec 2^n.3 \prec \cdots \\
&\cdots 11 \prec 9 \prec 7 \prec 5 \prec 3,
\end{aligned}$$

i.e., 1 followed by the powers of two ascending followed by ... followed by 2^{n+1} times the odds descending to three followed by 2^n times the odds descending to three and ending with the odds descending to three.

Theorem 2.2.12 (Sharkovskii). *Let $f: I \to \mathbb{R}$ be a continuous map of the interval I. If f has an orbit of least period p, then it has an orbit of least period m for all $m \prec p$ in the Sharkovskii order.*

A special case of Sharkovskii's Theorem is easy to prove using the techniques of Section 2.2.1, and we include this as an example of the power of the methods. A rather more detailed version of this result was proved by Li and Yorke in 1975 in a paper which includes the first use of the term 'chaos' [70].

Theorem 2.2.13. *Suppose $f: I \to \mathbb{R}$ is a continuous map of the interval and f has an orbit of period three. Then f has an orbit of period n for all $n \in \mathbb{Z}^+$ and f is chaotic.*

Proof. Let $x_1 < x_2 < x_3$ be the points on the orbit of period three. Then either

$$f(x_1) = x_2, \quad f(x_2) = x_3, \quad f(x_3) = x_1 \tag{2.13}$$

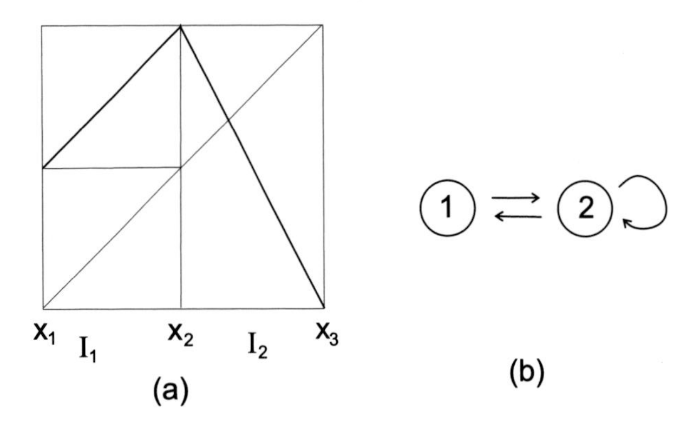

Figure 2.4: The order of the period three points and the intervals I_1 and I_2 with the transition graph.

or

$$f(x_1) = x_3, \quad f(x_2) = x_1, \quad f(x_3) = x_2. \tag{2.14}$$

These are equivalent under the reversal of x-axis, $x \to -x$, and so we will consider the first possibility (2.13) without loss of generality. Let

$$I_1 = [x_1, x_2], \quad I_2 = [x_2, x_3].$$

Now, since f is continuous $f(I_j)$ covers all points between the images of their end-points and so

$$I_2 \subseteq f(I_1) \quad \text{and} \quad I_1 \cup I_2 \subseteq f(I_2)$$

and the system has the associated transition graph of fig. 2.4. From Section 2.2.1, f has an orbit of period p for every closed loop in this transition graph. Hence it has a fixed point (I_2), an orbit of period two ($I_1 I_2$) and an orbit of period three by assumption. For all $n \geq 4$ it has a periodic orbit of period n by the closed loop $I_1 I_2^{n-1}$.

We leave it as an exercise to show from the transition graph that f^2 has a horseshoe and hence that f is chaotic by Definition 2.2.11. □

2.2.3 The square root map

Nordmark [78] introduced the square root map as a model for grazing in impacting systems. We shall choose the parametrization so that it is

$$S(x) = \begin{cases} \mu + bx & \text{if } x < 0, \\ \mu - \sqrt{x} & \text{if } x \geq 0, \end{cases} \qquad 0 < b < 1. \tag{2.15}$$

The map is continuous (and hence all the results from Section 2.2.2 hold) but it is not differentiable at $x = 0$. If $\mu < 0$ there is a stable fixed point in $x < 0$

which attracts all solutions so we will restrict attention to $\mu > 0$. As μ tends to zero, a sequence of stable periodic orbits can be observed numerically if b is small enough, with period two being followed by period three and then four and so on. This is an example of a period-adding sequence, although there is a region of parameters where the orbits of period n and $n+1$ coexist and also that as soon as the period three orbit exists, there are orbits of all periods by Sharkovskii's Theorem (Theorem 2.2.12), and so this does not give the whole picture of the dynamics.

In this section a simplified version of Nordmark's result will be proved relying only on the bifurcation structure of the simplest periodic orbits, those with one point in $x > 0$ and n in $x < 0$ (these are referred to as RL^n orbits for obvious reasons — one point to the right of the critical point and n to the left). In particular we will not prove that these are the only stable periodic orbits, though this follows from Nordmark's more detailed analysis [78].

The classic theory of period-doubling bifurcations [16] states that if x^* is a fixed point of a smooth map $x_{n+1} = f(x_n, \mu)$ at $\mu = \mu^*$ and $f_x(x^*, \mu^*) = -1$, then a period-doubling bifurcation occurs (creating a period two orbit) if

$$u = 2f_{\mu x} + f_\mu f_{xx} \neq 0, \quad v = \tfrac{1}{2}f_{xx}^2 + \tfrac{1}{3}f_{xxx}^2 \neq 0. \tag{2.16}$$

The bifurcating period two orbit is unstable if it coexists with the stable fixed point, which occurs if

$$v < 0 \tag{2.17}$$

in which case it is called a subcritical period-doubling bifurcation.

Theorem 2.2.14. *Consider the map S of (2.15). If $0 < b < \frac{1}{4}$, then there exist two sequences μ_n^{bc} and μ_n^{pd} converging on $\mu = -$ from above such that if $\mu = \mu_n^{bc}$ a stable periodic orbit RL^n is created in a border collision bifurcation at $x = 0$, whilst if $\mu = \mu_n^{pd}$ then this orbit loses stability by a classical subcritical period-doubling bifurcation. Moreover,*

$$\mu_n^{bc} > \mu_{n-1}^{pd} > \mu_{n+1}^{bc}. \tag{2.18}$$

Note that if $\mu_n^{bc} > \mu > \mu_{n-1}^{pd}$, then there are stable periodic orbits RL^{n-1} and RL^n, whilst if $\mu_{n-1}^{pd} > \mu > \mu_{n+1}^{bc}$, the RL^{n-1} orbit has lost stability and the only stable orbit is RL^n.

Proof. First note that if $x_k = S^k(x_0) < 0$, $k = 0, 1, 2 \dots, n-1$, then by induction or direct solution of the linear difference equation in $x < 0$,

$$S^n(x_0) = (1 + b + \cdots + b^{n-1})\mu + b^n x_0 = \frac{1 - b^n}{1 - b}\mu + b^n x_0. \tag{2.19}$$

A border collision bifurcation occurs if $x = 0$ lies on a periodic orbit. Now if $\mu > 0$ then $S(0) = \mu$ and $S^2(0) = \mu - \sqrt{\mu}$. Thus $S^{n+1}(0) = 0$ if $S^{n-1}(\mu - \sqrt{\mu}) = 0$ or, using (2.19),

$$0 = (1 + b + \cdots + b^{n-2})\mu + b^{n-1}(\mu - \sqrt{\mu}) = (1 + b + \cdots + b^{n-2} + b^{n-1})\mu - b^{n-1}\sqrt{\mu}.$$

Thus the border-collision bifurcation occurs for $\mu = \mu_n^{bc}$ where

$$\frac{1-b^n}{1-b}\sqrt{\mu} = b^{n-1} \quad \text{or} \quad \mu_n^{bc} = \frac{b^{2(n-1)}(1-b)^2}{(1-b^n)^2}. \tag{2.20}$$

The orbit will be stable provided the derivative along the orbit has modulus less than one at the bifurcation, i.e., provided

$$b^n\left(\frac{1}{2\sqrt{\mu}}\right) < 1,$$

and using (2.20) this condition becomes

$$b^n\frac{1-b^n}{b^{n-1}(1-b)} = \frac{b(1-b^n)}{1-b} < 1,$$

which is clearly satisfied if $b < \frac{1}{4}$.

There is a period-doubling bifurcation if the slope of S^{n+1} at the periodic point is -1, and hence if $x > 0$ is the point on the orbit to the right of $x = 0$, then for the RL^n orbit

$$(S^{n+1})'(x) = -b^n\frac{1}{2\sqrt{x}},$$

and so the periodic orbit RL^n period-doubles if the point of the orbit in $x > 0$ is

$$x = \frac{1}{4}b^{2n}. \tag{2.21}$$

If $x > 0$ is followed by n iterates in $x < 0$ and then closes to form a periodic orbit, then

$$x = S^{n+1}(x) = (1 + b + \cdots + b^{n-1})\mu + b^n(\mu - \sqrt{x}), \tag{2.22}$$

and for the choice of x in (2.21) for a period-doubling equation this becomes

$$\frac{3}{4}b^{2n} = (1 + b + \cdots + b^{n-1} + b^n)\mu,$$

so

$$\mu_n^{pd} = \frac{3b^{2n}(1-b)}{4(1-b^{n+1})}. \tag{2.23}$$

To check whether this is supercritical or subcritical we need to check the coefficients u and v defined in (2.16) with f replaced by S^{n+1} from the right-hand equality in (2.22). Thus

$$f_\mu = \tfrac{1-b^{n+1}}{1-b}, \quad f_x = -\tfrac{b^n}{2\sqrt{x}}, \quad f_{\mu x} = 0,$$
$$f_{xx} = \tfrac{b^n}{4}x^{-\frac{3}{2}}, \quad f_{xxx} = -\tfrac{3b^n}{8}x^{-\frac{5}{2}}.$$

Thus

$$u = \frac{1-b^{n+1}}{1-b}\frac{b^n}{4}x^{-\frac{3}{2}} > 0$$

and

$$v = \frac{1}{32}b^{2n}x^{-3} - \frac{1}{8}b^n x^{-\frac{5}{2}} = \frac{b^n x^{-3}}{32}\left(b^n - 4\sqrt{x}\right).$$

Substituting $\sqrt{x} = \frac{1}{2}b^n$ from (2.21) shows that the final term in brackets is $-b^n < 0$, and so $v < 0$ and by (2.17) the bifurcation is subcritical as stated.

Finally we need to establish (2.18). From (2.20) and (2.23) the second inequality $\mu_{n-1}^{pd} > \mu_{n+1}^{bc}$ is equivalent to

$$\frac{3b^{2(n-1)}(1-b)}{4(1-b^n)} > \frac{b^{2n}(1-b)^2}{(1-b^{n+1})^2}$$

or

$$\frac{3(1-b^{n+1})^2}{4(1-b^n)} > b^2(1-b),$$

which is clearly true if $0 < b < \frac{1}{4}$ as the right-hand side is greater than $\frac{3}{4}(1-b^{n+1}) > \frac{9}{16}$ whilst the left-hand side is less than $\frac{1}{16}$. The interesting inequality is the first of (2.18), $\mu_n^{bc} > \mu_{n-1}^{pd}$. From (2.20) and (2.23) this is equivalent to

$$\frac{b^{2(n-1)}(1-b)^2}{(1-b^n)^2} > \frac{3b^{2(n-1)}(1-b)}{4(1-b^n)}$$

or

$$1 - b > \frac{3}{4}(1-b^n),$$

i.e., $b - \frac{3}{4}b^n < \frac{1}{4}$, which is true for all $n \geq 1$ if $0 < b < \frac{1}{4}$. □

2.3 Piecewise-smooth maps of the interval

The next four sections describe properties of one-dimensional piecewise-smooth maps. In this section we describe some properties and analyse some simple examples. We need a technical convention about how to work with closed intervals if a map has a discontinuity.

Let $J = [a, b]$ be a closed interval and suppose that f is continuous on the interior of J. Then define

$$f(a) = \lim_{x \downarrow a} f(x) \quad \text{and} \quad f(b) = \lim_{x \uparrow b} f(x).$$

Note that when applying this to an iterate of J we may effectively be using two values of the map at the discontinuity. We will call this the closed set convention.

2.3.1 Transitivity and chaos

We start with a generalization of a horseshoe for piecewise-smooth maps which we will use to generalize the definition of chaos for continuous maps (definition 2.2.11) for maps which may have discontinuities.

Definition 2.3.1. Suppose $f: I \to I$ is a piecewise-smooth map of the interval I. f is chaotic if there exist disjoint open intervals J_0 and J_1 and $n_0, n_1 > 0$ such that $f^{n_k}|J_k$ is continuous and J_k f^{n_k}-covers J_0 and J_1.

Note that we are implictly using the closed set convention to extend the definition of covers to open sets. Two further definitions will be useful.

Definition 2.3.2. Suppose $f: I \to I$ is a piecewise-smooth map of the interval I. f is transitive if for every open interval $J \subseteq I$ there exists $N < \infty$ such that

$$I = \mathrm{c}\ell \bigcup_{k=0}^{N} f^k(U).$$

Note that this implies another definition of transitivity, that for all open intervals U and V in I there exists $k \geq 0$ such that $f^k(U) \cap V \neq \varnothing$. Moreover it can be used to prove that the non-wandering set of a map is the interval I itself.

Definition 2.3.3. Suppose $f: I \to I$ is a piecewise-smooth map of the interval I. A point is wandering if there exists an open set U with $x \in U$ such that $f^n(U) \cap U = \varnothing$ for all $n \geq 1$. If x is not a wandering point then x is a non-wandering point. The non-wandering set of f, $\Omega(f)$, is the set of all non-wandering points of f.

Lemma 2.3.4. *Suppose $f: I \to I$ is a piecewise-smooth map of the interval I. If f is transitive on I, then $\Omega(f) = I$.*

Proof. If f is transitive, then for any interval U there exists $N < \infty$ such that $I = \cup_0^N f^r(U)$ and hence $m \leq N$ such that $f^m(U) \cap U \neq \varnothing$; in other words, no point can be wandering. □

A stronger definition of the expansion of intervals makes all the consequences easy to establish.

Definition 2.3.5. Suppose $f: I \to I$ is a piecewise-smooth map of the interval I with M continuous, monotonic branches on the open intervals $J_1, \dots J_M$. Then f is locally eventually onto (LEO) if for every open interval $U \subseteq I$ there exist open intervals $L_k \subseteq U$ and $n_k \geq 0$ such that $f^{n_k}|L_k$ is monotonic and continuous and $f^{n_k}(L_k) = J_k$, $k = 1, \dots, M$.

Note that the conditions that $f^{n_k}|L_k$ are continuous imply that all the standard smooth dynamical results can be imported to the piecewise-smooth case provided a little care is taken with the end-points of the open intervals involved.

Lemma 2.3.6. *Suppose $f: I \to I$ is a piecewise-smooth map of the interval I. If f is LEO, then it is transitive and chaotic.*

Proof. Transitivity is obvious as for any open U there exist $L_k \subseteq U$, $k = 1, \dots, M$ as in the definition such that $\cup_1^M f^{n_k}(L_k) = \bigcup_1^M J_k$ and by definition the closure of the union of the monotonic branches is the whole interval.

For chaos (Definition 2.3.1), take two disjoint open intervals U and V in the same monotonic branch interval J_c. Then there exis $L_0 \subseteq U$ and $L_1 \subseteq V$ and $n_0, n_1 \geq 1$ such that $f^{n_k}|L_k$, $k = 0, 1$, is continuous and $f^{n_k}(L_k) = J_c$. Since $L_k \subset J_c$, f is chaotic by using the closed set convention at the beginning of Section 2.3 to extend to the closures of L_k. □

2.3.2 Tent maps

An interesting example is provided by the (symmetric) tent maps. This is a family of continuous piecewise-smooth maps of the interval $T_s\colon [0,1] \to [0,1]$ defined for $s \in (1,2]$ by

$$T_s(x) = \begin{cases} sx & \text{if } 0 \leq x \leq \frac{1}{2}, \\ s(1-x) & \text{if } \frac{1}{2} \leq x \leq 1. \end{cases} \tag{2.24}$$

Let the length of an interval U be denoted by $|U|$. There are three immediate remarks worth making to start with:

(a) there are no stable periodic orbits (the slope of the map has modulus $s > 1$);

(b) for every open interval $U \subset (0,1)$ there exists $n > 0$ such that $\frac{1}{2} \in T_s^n(U)$ (if not then T^s is linear so $|T_s(U)| = s|U|$ and by induction $|T_s^n(U)| = s^n|U|$; but since the interval must have length less than 1 this is a contradiction);

(c) let $I_0 = [T_s^2(\frac{1}{2}), T_s(\frac{1}{2}]$, then $\Omega(f) = \{0\} \cup \Omega(T_s|I_0)$ (0 is a fixed point so in $\Omega(T_s)$; the interval I_0 is invariant and any opn interval outside I_0 must map into I_0 eventually by (b)).

Lemma 2.3.7. *If* $\sqrt{2} < s \leq 2$ *and* I_0 *is as in* (c) *above, then* $\Omega(T_s) = \{0\} \cup I_0$. *These sets are disjoint unless* $s = 2$, *when* 0 *is the left end-point of* I_0.

Proof. We will show that $T_s|I_0$ is LEO and hence that $\Omega(T|I_0) = I_0$. By direct calculation $I_0 = [x_1, x_2]$ where

$$x_1 = \frac{s}{2}(2-s), \quad x_1 = \frac{s}{2}. \tag{2.25}$$

Note that if $s = 2$, then $I_0 = [0,1]$ and the last statement of Lemma 2.3.7 is shown.

Consider any open interval $U \subset I_0$. If $\frac{1}{2} \notin U$, then $|T_s(U)| = s|U|$ and so (cf. Remark (b) above) there exists n_0 such that $\frac{1}{2} \in T_s^{n_0}(U)$. Let $T_s^{n_0}(U) = V_0 \cup \{\frac{1}{2}\} \cup V_1$ with V_0 in $x < \frac{1}{2}$ and V_1 is in $x > \frac{1}{2}$. Then there exists $\alpha \in (0,1)$ such that

$$|V_0| = \alpha |T_s^{n_0}(U)|, \quad |V_1| = (1-\alpha)|T_s^{n_0}(U)|.$$

Both intervals $T_s(V_k)$ have $T_s(\frac{1}{2}) = x_2$ as their right end-point and so one contains the other (or both are equal). Thus

$$|T_s^{n_0+1}(U)| = \max_k\{|T_s(V_k)|\} = (\max\{\alpha s, (1-\alpha)s\})\,|T_s^{n_0}(U)|.$$

The maximum of αs and $(1-\alpha)s$ is greater than or equal to $\frac{1}{2}s$ since if $\alpha \neq \frac{1}{2}$, then one of the two terms α or $1-\alpha$ is greater than $\frac{1}{2}$. Hence

$$|T_s^{n_0+1}(U)| \geq \frac{s}{2}|T_s^{n_0}(U)|. \tag{2.26}$$

Choose $U_0 \subseteq U$ such that V_k where $T_s^{n_0}(U_0) = V_k$ is the interval with larger image, so $T_s^{n_0+1}|U_0$ is monotonic and $T_s^{n_0+1}(U_0) = T_s^{n_0+1}(U)$. If $\frac{1}{2} \notin T_s^{n_0+1}(U_0)$ then

$$|T_s^{n_0+2}(U_0)| \geq \frac{s^2}{2}|T_s^{n_0}(U_0)| > |T_s^{n_0}(U)| \tag{2.27}$$

and so the length continues to expand. This cannot continue indefinitely so after a finite number of further passages including $\frac{1}{2}$ (at which we define smaller intervals $U_1, U_2, \dots, U_m$ in U such that $|T_s^{n_r+1}(U_r)| = |T_s^{n_r+1}(U)|$ and $T_s^{n_r+1}|U_r$ is monotonic) we arrive at an interval U_m such that

$$\frac{1}{2} \in T_s^{n_m+1}(U_m) \quad \text{and} \quad \frac{1}{2} \in T_s^{n_m+2}(U_m).$$

But the first of these implies that $T_s(\frac{1}{2}) \subseteq T_s^{n_m+2}(U_m)$, so U_m contains an open interval $\tilde{U}$ such that $T_s^{n_m+2}|\tilde{U}$ is monotonic and $T_s^{n_m+2}(\tilde{U}) = (\frac{1}{2}, \frac{s}{2})$. Thus $T_s^{n_m+3}|\tilde{U}$ is monotonic and $T_s^{n_m+3}(\tilde{U}) = (\frac{s}{2}(2-s), \frac{s}{2})$ and so T_s is LEO on I_0. □

The next step is probably the most important in this course: it involves looking at a higher iterate of T_s on a subinterval of $[0,1]$. This is the idea behind renormalization and induced maps. We will make this explicit in the next subsection, but for the moment we will see it in action as we extend Lemma 2.3.7 to $1 < s \leq \sqrt{2}$.

Theorem 2.3.8. *If $\sqrt{2} < s^{2^n} \leq 2$, $n \geq 0$, then*

$$\Omega(T_s) = \{0\} \cup I_n \cup \left(\bigcup_{k=1}^{n} P_k\right)$$

where the right union is empty if $n = 0$. The set P_k is an unstable periodic orbit of period 2^k, $k = 1, 2, \dots, n$, and I_n is a union of 2^n closed intervals. These intervals are disjoint unless $s^{2^n} = 2$ in which case they intersect pairwise on the periodic orbit P_n.

Proof. If $n = 0$, the theorem is proved by Lemma 2.3.7. If $s \leq \sqrt{2}$, consider the second iterate of the map, T_s^2 which has the form shown in fig. 2.5. It has turning points at $\frac{1}{2}$ and the two preimages of $\frac{1}{2}$, i.e., $c_\pm$ where $sc_- = \frac{1}{2}$ and $s(1-c_+) = \frac{1}{2}$, solving gives

$$c_- = \frac{1}{2s}, \quad c_+ = \frac{2s-1}{2s}.$$

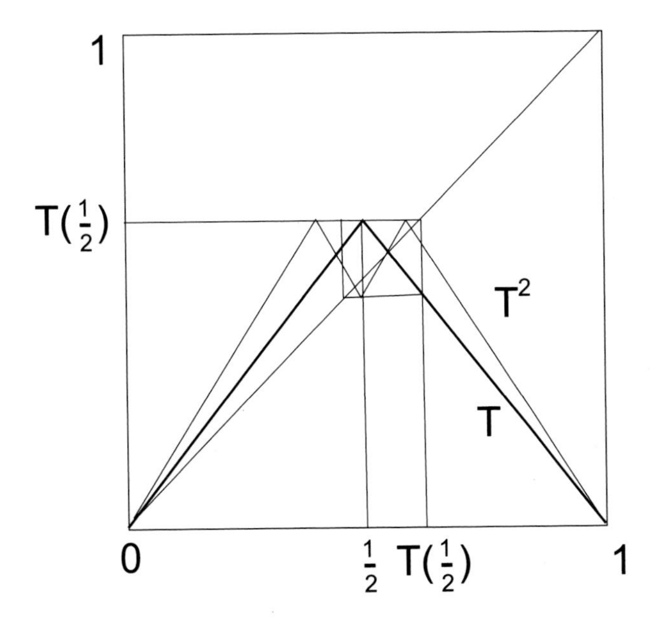

Figure 2.5: $s < \sqrt{2}$: the tent map and its second iterate.

There is a non-trivial fixed point of T_s in $x > \frac{1}{2}$ with at $x_* = \frac{s}{s+1}$ and this has a preimage in $x < \frac{1}{2}$, y_-, where $sy_- = x_*$, and this in turn has a preimage in $x > \frac{1}{2}$, y_+, with $s(1 - y_+) = y_-$. Direct calculation yields

$$y_- = \frac{1}{s+1} = 1 - \frac{s}{s+1}, \qquad y_+ = \frac{s^2 + s - 1}{s(s+1)}. \tag{2.28}$$

Consider $T_s^2|[y_-, x_*]$. This is symmetric about $\frac{1}{2}$ and the modulus of the slope is s^2. T_s^2 maps the interval $[y_-, x_+]$ into itself provided $T_s^2(\frac{1}{2}) \geq y_-$ which is equivalent to $s^2 \leq 2$ after some algebra. Thus if $s^2 \leq 2$, the map $T_s^2|[y_-, x_*]$ is equivalent by an affine change of variable to $T_{s^2}|[0, 1]$.

Thus if $\sqrt{2} < s^2 \leq 2$, $\Omega(T_s^2|[y_-, x_*]) = \{x_*\} \cup J_0$ where J_0 is an interval disjoint from x_* except in the case $s^2 = 2$ when x_* is an endpoint of J_0. Let $I_1 = J_0 \cup T_s(J_0)$ and $P_1 = \{x_*\}$. Then this establishes

$$\Omega(T_s) = \{0\} \cup I_1 \cup P_1, \quad \sqrt{2} < s^2 \leq 2,$$

where I_1 is a union of two intervals joined pairwise on P_1 if $s^2 = 2$.

To complete the proof use induction on n. If $s^2 \leq \sqrt{2}$, then consider the second iterate of T_s^2 on $[y_-, x_*]$, which has slopes of modulus $s^4 = s^{2^2}$ and the same structure provided $\sqrt{2} < s^4 \leq 2$, and so on. We leave the details to the reader. □

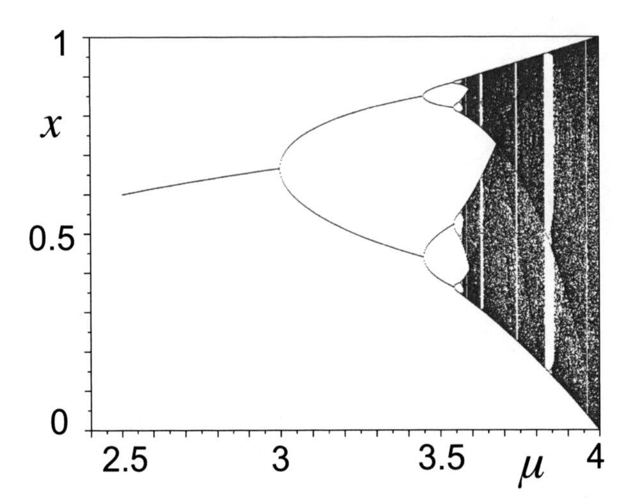

Figure 2.6: The attractor of the logistic map as a function of parameters.

2.3.3 Renormalization and period-doubling

The idea of looking at the second iterate of a map is the simplest way of looking at the standard phenomenon of period-doubling and the structure of smooth unimodal (or one-hump) maps such as the logistic family

$$x_{n+1} = \mu x(1-x), \quad 0 < \mu \leq 4, \quad x \in [0,1]. \tag{2.29}$$

For the logistic map the choice of $\mu \in (0,4]$ ensures that the interval $[0,1]$ is mapped into itself; if $\mu \in (0,1)$, then a simple argument based on Lemma 2.2.2 implies that all orbits tend to the fixed point at $x = 0$. Above this value a new fixed point, $x^* = (\mu - 1)/\mu$, appears and it is initially stable. It loses stability at μ_1 by a period-doubling bifurcation creating a stable period two orbit which in turn period-doubles at $\mu = \mu_2$. It is not obvious, but it is true that this cascade of period-doubling keeps going, accumulating at some value μ_∞. If $\mu_\infty < \mu \leq 4$, then there are parameter values at which there appear to be strange attractors, but also 'windows' of parameters which have stable periodic orbits. These orbits appear to have their own period-doubling sequences and the attracting behaviour as a function of parameter is shown in fig. 2.6. Understanding this picture and the existence of strange attractors for a positive measure set (containing no open intervals) of parameters was one of the achievements of dynamical systems theory of the 1980s.

An intuitive explanation of what is happening can be useful.

If $\mu = \mu_0 = 1$, then the fixed point at the origin has $f'(0) = 1$ and $f(\frac{1}{2}) = \frac{1}{4}$, so $f(x) < \frac{1}{2}$ for all $x \in [0,1]$ and $f^n(x) < f^{n-1}(x)$ for all $n > 1$, implying that all solutions tend to zero (even though it is not linearly stable).

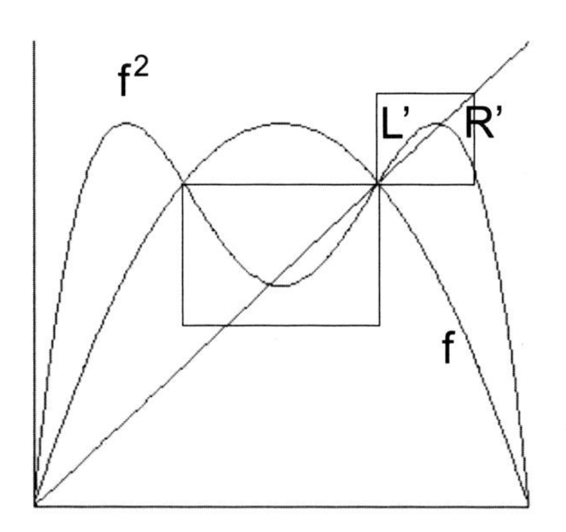

Figure 2.7: The map f and its second iterate f^2 if $2 < \mu < \tilde{\mu}_1$.

If $\mu = \tilde{\mu}_0 = 4$, then $f(\frac{1}{2}) = 1$ and so $f([0,\frac{1}{2}]) = [0,1] = f([\frac{1}{2},1])$. Thus f has a horseshoe and so it is chaotic.

If $\mu > 2$, the nontrivial fixed point

$$x^* = (\mu - 1)/\mu$$

lies in $x > \frac{1}{2}$ where the derivative is negative. In this case the *second* iterate of f, f^2 will look something like the sketch in fig. 2.7, where the two boxes are based on the fixed point, it's preimage y_- in $x < \frac{1}{2}$ and the preimage of that point, y_+, in $x > \frac{1}{2}$ (just like in the construction of T_s^2 in the previous section).

Now, if $\mu = \tilde{\mu}_0$, the $f^2(\frac{1}{2}) = 0 < y_-$ and so the central box is not mapped into itself (unlike the case sketched in fig. 2.7). Thus (by continuity) there exists $\tilde{\mu}_1$ such that at this value of the parameter $f^2(\frac{1}{2}) = y_-$ and if $\mu > \tilde{\mu}_1$ f^2 has a horseshoe, whilst if $2 < \mu < \tilde{\mu}_1$, the boxes $I_0 = [y_-, x^*]$ and $I_1 = [x^*, y_+]$ are mapped into themselves by the second iterate and permuted by the first iterate of f.

As μ increases from 2, there exists μ_1 at which $f'(x^*) = -1$ and the first period-doubling bifurcation occurs, with $(f^2)'(x^*) = [f'(x^*)]^2 = 1$ using the chain rule and $f(x^*) = x^*$.

Thus as μ varies between μ_1 and $\tilde{\mu}_1$, f^2 maps each of the intervals I_0 or I_1 into themselves and evolves essentially in the same way (see note below) as f for $\mu \in [\mu_0, \tilde{\mu}_0]$. If $\mu \in (\mu_0, \mu_1)$, then f has a stable fixed point whilst if $\mu \in (\tilde{\mu}_1, \tilde{\mu}_0)$, the two intervals are not mapped into themselves by f^2.

Hence there exist $\mu_2 < \tilde{\mu}_2$ in $(\mu_1, \tilde{\mu}_1)$ which play the same role for f^2 as μ_1 and $\tilde{\mu}_1$ do for f. In particular,

(i) if $\mu_1 < \mu < \mu_2$, then f^2 has a stable fixed point in each of the intervals I_0 and I_1; a stable orbit of period two for f;

(ii) if $\tilde{\mu}_2 < \mu < \tilde{\mu}_1$, then f^2 maps the intervals I_0 and I_1 into themselves and are permuted by f, i.e., any attractor lies in a union of two intervals ('two bands');

(iii) if $\mu_2 < \mu < \tilde{\mu}_2$, the second iterate of the second iterate, $f^4 = f^{2^2}$ evolves on two intervals in each of I_0 and I_1 as f^2 does on $\mu_1 < \mu < \tilde{\mu}_1$ which is the same as how f evolves on $\mu_0 < \mu < \tilde{\mu}_0$.

This self-similarity in the structure of the dynamics as a function of parameters continues, yielding nested intervals $[\mu_{n+1}, \tilde{\mu}_{n+1}] \subset [\mu_n, \tilde{\mu}_n]$ with $\lim_{n\to\infty} \mu_n = \lim_{n\to\infty} \tilde{\mu}_n = \mu_\infty$, which is called the accumulation of period-doubling. Moreover,

(iv) if $\mu_n < \mu < \mu_{n+1}$, then f^{2^n} has a stable fixed point in each of 2^n intervals; a stable orbit of period 2^n for f;

(v) if $\tilde{\mu}_{n+1} < \mu < \tilde{\mu}_n$, then there are 2^n intervals mapped into themselves by f^{2^n} and permuted by f, i.e., any attractor lies in a union of 2^n intervals or bands;

(vi) if $\mu_{n+1} < \mu < \tilde{\mu}_{n+1}$, the second iterate of the 2^nth iterate, $f^{2^{n+1}}$ evolves on 2^{n+1} intervals as f^{2^n} does on 2^n intervals for $\mu_n < \mu < \tilde{\mu}_n$, as ..., as f^2 does on $\mu_1 < \mu < \tilde{\mu}_1$ which is the same as how f evolves on $\mu_0 < \mu < \tilde{\mu}_0$.

Important note: *This argument rests on the assumption that all possible behaviour consistent with the constraints of a one-hump map actually does occur in this family of maps. The proof of this (what Collet and Eckmann* [16] *call a full family) relies on the family being C^1 and for both f and f' to be close for nearby values of μ; thus it does NOT necessarily hold for piecewise-smooth systems (though there are examples where it does hold).*

A further feature of this period-doubling cascade is notable: the convergence rate of μ_n and $\tilde{\mu}_n$ is independent of the details of the map: for one-hump maps with quadratic maxima or minima

$$\lim_{n\to\infty} \frac{\mu_{n-1} - \mu_n}{\mu_n - \mu_{n+1}} = \lim_{n\to\infty} \frac{\tilde{\mu}_{n-1} - \tilde{\mu}_n}{\tilde{\mu}_n - \tilde{\mu}_{n+1}} = \delta \tag{2.30}$$

where δ is the Feigenbaum constant, $\delta \approx 4.669$. For piecewise-smooth maps with maxima or minima of the form $|x|^r$, $r > 1$, a similar scaling holds but the constant becomes a function of r.

The explanation of Feigenbaum's result is nice. At μ_∞ the renormalization process (looking at the second iterate restricted to a subinterval) can be repeated infinitely often. This means that after rescaling and shifting so that the turning point is at $x = 0$ and $f(0) = 1$ then the rescaled map is

$$-\frac{1}{\alpha} f \circ f(-\alpha x), \quad \alpha = -f(1) > 0.$$

This suggests viewing the process as a map in function space, $f \to \mathcal{T} f$ where

$$\mathcal{T} f = -\frac{1}{\alpha} f \circ f(-\alpha x). \tag{2.31}$$

It turns out that $\mathcal{T}$, appropriately defined, has a fixed point f_* with a one-dimensional unstable eigenspace with eigenvalue δ. Functions on the stable manifold of f_* tend to f_* under iteration and can be renormalized infinitely often, i.e., they are at the accumulation of period-doubling. Codimension one surfaces of period-doubling bifurcations converge one the stable manifold under $\mathcal{T}^{-1}$ at a rate determined by δ^{-1} explaining (at least up to the many technical details that have been avoided in this brief description) the exponential convergence (2.30).

One further feature of the period-doubling phenomenon is worth considering, particularly as it will be useful in the study of an example later, and a similar argument will be used in the next section when considering circle maps. Orbits that avoid the turning point can be labelled by sequences of Rs and Ls according to whether the nth iterate is on the left (L) or the right (R) of the turning point. At the accumulation of period-doubling the sequence for $f(c)$, where c is a maximum, is

$$RLRRRLRLRLRRRLRRRLRRRLRLRLRLRRRLRLR\cdots. \tag{2.32}$$

This sequence can be generated by looking at fig. 2.7. The one-hump map in the interval I_1 can be described by symbols R' and L'; these are the symbols for f^2 restricted to the right-hand interval I_1, and for f the symbol R' actually means RL (it is a decreasing branch of f^2 on the right, so $f(x) < c$ here); and L' for f^2 means RR for f. Thus the replacement operations

$$L \to RR, \quad R \to RL \tag{2.33}$$

can be used to translate from f^2 to f.

But this means that for f^4 the replacement operation (2.33) to obtain sequences in terms of the original function f is obtained by repeating twice, i.e.,

$$L \to RR \to RLRL, \quad R \to RL \to RLRR.$$

Moreover, $RLRR$ is the symbol sequence corresponding to the fixed point with code R for f^4, i.e., the period-doubled orbit of the period-doubled orbit for f. At the accumulation of period-doubling the process repeats infinitely often and applying this to the rightmost point at every stage we find (2.32).

Lemma 2.3.9. *The number of Rs in the sequence associated with the period-doubling orbit of period 2^n is*

$$r_n = \tfrac{2}{3}2^n + \tfrac{1}{3}(-1)^n, \tag{2.34}$$

and the asymptotic proportion of Rs at the accumulation of period-doubling is $\frac{2}{3}$.

Proof. Let (r_n, ℓ_n) denote the number of Rs and Ls in the orbit of period 2^n with code R for f^{2^n}. Then the code for the orbit of period 2^{n+1} is obtained by replacing each of the r_n Rs by RL and each of the ℓ_n Ls by RR using (2.33), so

$$r_{n+1} = r_n + 2\ell_n, \quad \ell_{n+1} = r_n$$

and so

$$r_{n+1} = r_n + 2r_{n-1}. \tag{2.35}$$

This is a linear difference equation and so if we pose solutions of the form s^n, we find $s^2 - s - 2 = 0$ and so $s = 2$ or $s = -1$, and the general solution of (2.35) is

$$r_n = A2^n + B(-1)^n, \tag{2.36}$$

for some constants A and B. Since the first two sequences are R (the fixed point) and RL (period two), the initial conditions are $r_0 = 1$, $r_1 = 1$, i.e., $A + B = 1$, $2A - B = 1$, which imply that $A = \frac{2}{3}$ and $B = \frac{1}{3}$ proving (2.34). The asymptotic proportion of Rs is obtained by taking the limit of $\frac{r_n}{2^n}$ which is $\frac{2}{3}$. □

For completeness we should note that

$$\ell_n = r_{n-1} = \frac{1}{3}(2^n - (-1)^n). \tag{2.37}$$

The proof of this lemma was one of the exercises set during the Advanced Course in Barcelona, 2016.

2.3.4 Renormalization and induced maps

The previous example is our first sight of a really important idea: renormalization, i.e., the consideration of induced maps. This will be central to much of the theoretical analysis we do here and can be used to provide a detailed description of the dynamics of piecewise-smooth maps.

Definition 2.3.10. Suppose $f: [0,1] \to [0,1]$ is a piecewise-smooth map and there exists $c \in (0,1)$ such that f is monotonic and continuous on $(0,c)$ and on $(c,1)$. f is renormalizable if there exist positive integers n_0 and n_1 with $n_0 + n_1 > 2$ and non-trivial intervals $J_0 = (x_1, c)$ and $J_1 = (c, x_2)$ such that $f^{n_k}|J_k$ is continuous and monotonic, $k = 0,1$ and

$$f^{n_k}(J_k) \subseteq J_0 \cup \{c\} \cup J_1, \quad k = 0,1. \tag{2.38}$$

In some sense, apart from stable periodic orbits, renormalization is the only obstruction to transitivity in piecewise-smooth maps with two monotonic branches.

Theorem 2.3.11. *Suppose $f: [0,1] \to [0,1]$ is a piecewise-smooth map with two monotonic branches separated by $c \in (0,1)$. If there exists $s > 1$ such that $|f'(x)| \geq a$ for all $x \in (0,1) \backslash \{c\}$, then either f is transitive or f is renormalizable. If f is renormalizable on an interval J containing c, then $\Omega(f) = T \cup R$ where T is described by a Markov graph and R is the nonwandering set of the induced map on J and its iterates under f.*

Proof. Without loss of generality assume that $[0,1]$ is the smallest interval mapped into itself by f. Note that f has no stable periodic orbits.

Take any open interval U. By the expansion argument of (b) above Lemma 2.3.7 there exists $n_0 \geq 0$ such that $c \in f^{n_0}(U) = U_0$ and follow both branches to their next intersection with c, i.e., let $U_0 = V_0 \cup \{c\} \cup V_1$ in the standard way and choose the smallest m_k, $k = 0, 1$ such that $c \in f^{m_k}(V_k)$, $k = 0, 1$ (these exist by the expansion argument). If $f^{m_k}(V_k) \subseteq U_0$, then f is renormalizable. Otherwise set $U_1 = f^{m_0}(V_0) \cup f^{m_1}(V_1) \cup U_0$ and note $U_0 \subset U_1$.

Now repeat the argument using U_1 and note that the equivalent of the return times for U_1 are less than or equal to the return times m_k for U_0. Either f is renormalizable or there exists U_2 with $U_1 \subseteq U_2$ which is a union of iterates of subsets of U.

Either there exists $m < \infty$ such that $U_m = (0,1)$ and so U satisfies the transitivity condition (but not necessarily all U satisfy the condition) or $U_n \to U_\infty$ as $n \to \infty$ and by continuity appropriate iterates of f map U_∞ into itself.

Hence once again either f is renormalizable or $U_\infty = (0,1)$. But since the return times are decreasing, they also tend to a limit, m_k^∞, $k = 0, 1$, and these are reached in finite steps. Thus if $U_n \neq (0,1)$ for all $n > N_0$ the minimality of $(0,1)$ implies that $U_n \cup f(U_n) = (0,1)$ for large enough n; the transitivity condition again.

Thus for each U either U satisfies the transitivity condition or f is renormalizable. Hence either f is renormalizable or f is not renormalizable and every open U satisfies the transitivity condition and hence f is transitive.

If f is renormalizable, let J_k be the intervals as in the definition and choose the maximal intervals satisfying (2.38). Let

$$K = J_0 \cup \left(\bigcup_1^{n_0} f^r(J_0) \right) \cup \{c\} \cup J_1 \cup \left(\bigcup_1^{n_1} f^r(J_1) \right)$$

and let $L = I \backslash K$. Then L is a (possibly empty) finite union of closed intervals and since the sets K are mapped to themselves if $f(L_i) \cap L_j \neq \varnothing$ then $L_j \subseteq f(L_i)$, i.e., L_i f-covers L_j and so the dynamics in L can be described by a Markov graph. Setting $T = \Omega(f) \cap L$ and $R = \Omega(f) \cap c\ell(K)$ produces the stated decomposition of the non-wandering set. □

Whilst this result provides a clear description of the possible dynamics of these maps, a description that can be extended to maps without uniform expansion using the ideas of [65, 73] and which can be made unique by choosing the shortest renormalization (smallest $n_0 + n_1$ at each stage), it does not describe how to determine which of the many possibilities actually occurs in a given family. This can be an extremely complicated question to answer in detail, and in the spirit of 'less is more' we believe that this level of general description is more useful than an exhaustive description of cases in a particular family (which will change with the details of the family) except in cases where that family has particular significance. Dan Berry [11] provides a detailed description of the decomposition of the non-wandering set for piecewise-smooth maps with two monotonic branches.

2.3.5 Boundary bifurcations

In the previous sections we have been concerned with chaos and expansion. Now we consider how periodic orbits can be created or destroyed by non-smooth effects. To do bifurcation theory we need to consider families of maps, and this leads to problems about how to talk about 'continuous' families of discontinuous mappings! In the next section we will look at some more sophisticated approaches, but for now we are concerned only with local phenomena and so we will work with locally fixed families.

Definition 2.3.12. A family of piecewise-smooth mappings $f(x,\mu)$, $f:[0,1]\times\mathbb{R}\to[0,1]$ is locally fixed if there exists $\epsilon>0$ such that the set of discontinuities, d_k, and the set of critical points, c_k are fixed for all $\mu\in(-\epsilon,\epsilon)$ and f is C^2 functions of both variables on the intervals J_k.

Thus for a locally fixed family, there exist fixed intervals bounded by the discontinuities and critical points on which f is smooth. For most families this can be achieved locally by a change of coordinates.

Theorem 2.3.13. *Let $f:I\times(-\epsilon,\epsilon)\to I$ be a locally fixed family of piecewise-smooth maps and suppose that d is a point of discontinuity. If there is a neighbourhood $J=(d,d+\delta)$, $\delta>0$, such that f is smooth (C^2),*

$$\lim_{x\downarrow d} f(x,0)=d,\quad \lim_{x\downarrow d}|f'(x,0)|=a\neq 1, \tag{2.39}$$

then there exists $\delta,\eta>0$ such and $b\in\{+1,-1\}$ such that if μ is between 0 and $b\eta$, then f has a fixed point in $(d,d+\delta)$ and no other locally recurrent dynamics, whilst if μ is between 0 and $-b\eta$, then f has no locally recurrent dynamics. The fixed point is stable if $a<1$ and unstable if $a>1$.

Thus the effect of a boundary bifurcation is to create or destroy a fixed point. Of course the same result holds for periodic orbits by replacing f by f^p. Global features of the maps can create more dynamics in the intervals — the theorem only refers to dynamics locally, i.e., that remain in the interval J for all time. The proof is elementary and is left as an exercise. The four cases are illustrated in fig. 2.8.

2.3.6 Topological conjugacy and unbounded domains

It should be obvious that if two systems are related by a change of coordinate then we need not analyze them separately — understanding the dynamics of one of the systems is enough to describe the dynamics of both systems after translating

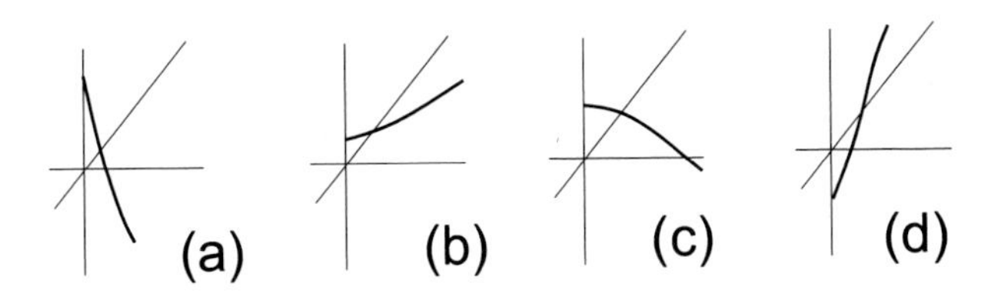

Figure 2.8: The four cases for elementary boundary bifurcations: (a) $a < -1$; (b) $0 < a < 1$; (c) $-1 < a < 0$; (d) $a > 1$.

between systems using the change of coordinates. Thus given $x_{n+1} = f(x_n)$ and an invertible change of coordinates $y = h(x)$,

$$y_{n+1} = h(x_{n+1}) = h(f(x_n)) = h(f(h^{-1}(y_n))).$$

Thus the dynamics of y is determined by a new map g where $g = h \circ f \circ h^{-1}$. This idea is made formal by the definition of a topological conjugacy.

Definition 2.3.14. The maps $f: I \to I$ and $g: J \to J$ are topologically conjugate if there exists a homeomorphism (continuous with continuous inverse) $h: I \to J$ such that

$$g \circ h = h \circ f. \tag{2.40}$$

This is often described by saying that the following diagram commutes.

$$\begin{array}{ccc} I & \xrightarrow{f} & I \\ {\scriptstyle h}\downarrow & & \downarrow{\scriptstyle h} \\ J & \xrightarrow{g} & J \end{array}$$

As a simple example consider the tent map $T_s: [0,1] \to [0,1]$

$$T_s(x) = \begin{cases} sx & \text{if } x \in [0, \frac{1}{2}], \\ s(1-x) & \text{if } x \in [\frac{1}{2}, 1]. \end{cases}$$

These could equally be rescaled to maps on the interval $[-1,1]$. To see how to do this note the symmetry of T_s about the mid-point of the interval $[0,1]$ so choose a transformation that takes 0, $\frac{1}{2}$, 1 to -1, 0, 1. The easiest example of such a transformation is linear:

$$y = h(x) = 2x - 1.$$

In these new coordinates $y_{n+1} = 2x_{n+1} - 1$, i.e., $y_{n+1} = g_s(y_n)$ where

$$g_s(y) = \begin{cases} sy + s - 1 & \text{if } y \in [-1, 0], \\ -sy + s - 1 & \text{if } y \in [0, 1]. \end{cases}$$

Clearly this is just a change of scale; the alternative $y = 1 - 2x$ makes the function a minimum instead of a maximum.

We have worked on finite intervals and there are some technical issues about unbounded domains which we will ignore, but Gardini et al [39] have recently introduced some piecewise-smooth unbounded maps which have nice properties that can be studied using the techniques introduced here.

One of the examples in [39] is

$$G_1(x) = \begin{cases} \mu + bx & \text{if } x \in (-\infty, 0], \\ \mu - cx^{-\gamma} & \text{if } x \in (0, \infty), \end{cases} \tag{2.41}$$

with

$$0 < b < 1, \quad c > 0, \quad \gamma > 0, \tag{2.42}$$

and $\mu > 0$ as a parameter. Note that these inequalities imply that

$$G_1(x) \to -\infty \text{ as } x \downarrow 0, \quad \text{and} \quad G_1(x) \leq \mu.$$

Hence we can think of G_1 as acting on $(-\infty, \mu]$. Gardini et al [39] comment on the existence of unbounded chaos in this example, and it is natural to ask what the role of the unbounded domain might be. This suggests looking for topologically conjugate systems on finite domains, and a natural change of variable is a (real) Möbius transformation that brings

$$\infty \to -1, \quad 0 \to 0, \quad \mu \to \mu, \tag{2.43}$$

although it would have been just as natural to choose $\mu \to 1$ so that the dynamics is on the interval $[-1, 1]$ rather than $[-1, \mu]$.

The Möbius transform that achieves this is

$$h(x) = \frac{x}{1 + \mu - x}$$

which is continuous for $x \in (-\infty, 1 + \mu)$ and hence on $(-\infty, \mu)$ and

$$h'(x) = \frac{1 + \mu - x - (-x)}{(1 + \mu - x)^2} = \frac{1 + \mu}{(1 + \mu - x)^2} > 0$$

so h is strictly increasing and differentiable on $(-\infty, \mu]$ and hence invertible (but not a diffeomorphism as the derivative tends to zero as x tends to $-\infty$). Let

$$g_1 = h \circ G_1 \circ h^{-1}$$

then,

(i) $g_1(-1) = -1$;

(ii) $g_1(0^-) = \mu$;

(iii) g_1 is strictly increasing on $(-1,0)$ and $(0,\mu)$;

(iv) $g_1(x) > x$ on $(-1,0)$;

(v) $g_1(0^+) = -1$.

Lemma 2.3.15. $g_1: [-1,\mu] \to [-1,\mu]$ *is chaotic.*

Proof. We have $g_1((-1,0)) = (-1,\mu)$ and there exists $n \geq 0$ such that $g_1^n(\mu) > 0$ and g_1^n restricted to $(0,\mu)$ is continuous by (iv). Hence $(-1,0) \subset g_1^n((0,\mu))$ and so there exists $J \subseteq (0,\mu)$ such that $g_1^{n+1}(J) = (-1,\mu)$ and g_1^{n+1} is continuous on J. Hence g_1 is chaotic by Definition 2.3.1. □

2.4 Lorenz maps and rotations

This section is devoted to piecewise-smooth maps of the interval with a single discontinuity such that both continuous branches are increasing. These include Lorenz maps and rotations. If $f: [0,1] \to [0,1]$ is a piecewise-smooth map with increasing branches and a single discontinuity at $c \in (0,1)$ with $f(c_-) = 1$ and $f(c_+) = 0$ there are three separate cases:

(i) rotations: $f(0) = f(1)$,

(ii) gap maps: $f(0) > f(1)$,

(iii) overlap maps: $f(0) < f(1)$.

Rotations have a distinguished history going back to the classic results of Julia and Denjoy in the early twentieth century. Gap maps have many similarities and some beautiful general results are due to Keener [66] and Rhodes–Thompson [83, 84]. Overlap maps allow the possibility of chaos and include the many studies of Lorenz maps.

2.4.1 Rotations

A rigid rotation is a map $r_\alpha: [0,1) \to [0,1)$ with $\alpha \in [0,1)$ defined by

$$r_\alpha(x) = x + \alpha \pmod 1. \tag{2.44}$$

The function $R_\alpha: \mathbb{R} \to \mathbb{R}$ defined by $R_\alpha(x) = x + \alpha$ is an example of a lift of r_α, and $R_\alpha(x+1) = R_\alpha(x) + 1$. The dynamics of the map r_α can be recovered from R_α by projecting modulo 1, hence r_α has a periodic point of period q iff there exists $x \in \mathbb{R}$ such that $R_\alpha^q(x) = x + p$ for some $p \in \mathbb{Z}$ (so $x + p = x$ mod 1). But

$$R_\alpha^q(x) = x + q\alpha \tag{2.45}$$

so x is periodic if and only if $q\alpha = p$, or $\alpha = \frac{p}{q} \in \mathbb{Q}$, and in this case all points are periodic. If $\alpha \notin \mathbb{Q}$, then the orbit is dense on the circle (see, e.g., Devaney [18]). Thus for rigid rotations there is a simple dichotomy

(i) $\alpha \in \mathbb{Q}$ and all points are periodic; or

(ii) $\alpha \notin \mathbb{Q}$ and orbits are dense on the circle.

Note also that (2.44) can be seen as a map of the interval with a discontinuity: given $\alpha \in (0,1)$ define $f_\alpha: [0,1] \to [0,1]$ by

$$f_\alpha(x) = \begin{cases} x + \alpha & \text{if } 0 \le x < 1 - \alpha, \\ x + \alpha - 1 & \text{if } 1 - \alpha < x < 1, \end{cases} \tag{2.46}$$

with our usual convention about the discontinuity. We will exploit this connection more in the next section, but first we describe some of the classic results for orientation preserving homeomorphisms of the circle.

The generalization of α to orientation preserving homeomorphisms is the idea of a rotation number which describes an average angular velocity around the circle.

Definition 2.4.1. If f is a circle map with lift F, then, provided the limit exists,

$$\rho(F,x) = \lim_{n\to\infty} \tfrac{1}{n}(F^n(x) - x) \tag{2.47}$$

is called the rotation number of x under F.

The following sequence of theorems describes the classic results of Julia and Denjoy. Proofs use simple real analysis and can be found in Devaney [18].

Theorem 2.4.2. *If F is the lift of an orientation preserving homeomorphism of the circle f, then $\rho(F,x)$ exists and is independent of x.*

It is usual to talk about the rotation number of f in this case, denoted $\rho(f)$ as $\rho(F,x)$ modulo 1.

Theorem 2.4.3. *Suppose f is an orientation preserving homeomorphism of the circle.*

(i) *If $\rho(f) \in \mathbb{Q}$, then f has at least one periodic orbit.*

(ii) *If $\rho(f) \notin \mathbb{Q}$, then f has no periodic orbits and if f is C^2 then every orbit is dense in the circle.*

If f is not C^2, then it is possible to create attracting Cantor sets with irrational rotation numbers, these are the Denjoy counter-examples.

Families of circle maps can be defined via their lifts: a continuous family of smooth circle maps is a family with lifts F_μ which can be chosen such that such that

$$\lim_{\mu\to\mu_0} |F_\mu(x) - F_{\mu_0}(x)| = 0$$

for all $x \in \mathbb{R}$.

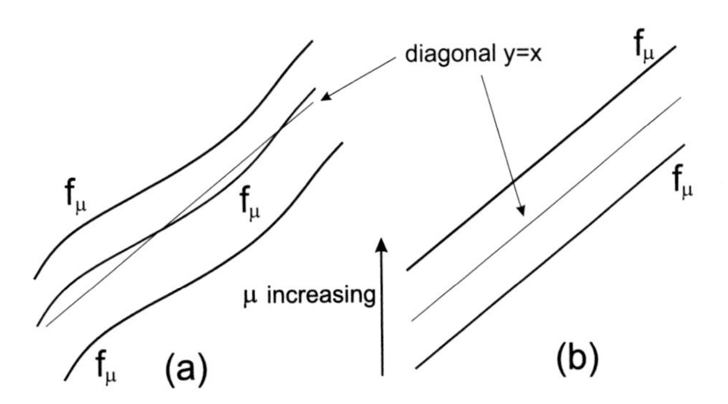

Figure 2.9: Part of the graph of families of maps f_μ as μ increases illustrating why there is (a) persistency of intersections with the diagonal in the general case as the graph needs to be shifted significantly to avoid intersections with the diagonal; and (b) degeneracy in the slope one case.

Theorem 2.4.4. *If (f_μ) is a continuous family of continuous circle orientation preserving homeomorphisms, then $\rho(f_\mu) = \rho(\mu)$ varies continuously. If there exist $\mu_1 < \mu_2$ such that $\rho(\mu_1) < \frac{p}{q} < \rho(\mu_2)$ then typically $\rho(\mu) = \frac{p}{q}$ on an interval of parameter values.*

Again, we will not give proofs for the circle maps case; see [18]. The continuous variation of $\rho(\mu)$ implies that the irrational rotation numbers do appear in examples. The interval of values with rational rotation numbers is often referred to as mode locking. It is easy to see why this occurs typically. If $\rho(\mu) = p/q$, then, by Theorem 2.4.3(i), there exists a periodic point, i.e., a solution to $F^q(x) = x + p$. If $\rho(\mu) < \frac{p}{q}$, then $F^q(x) < x+p$ for all $x \in \mathbb{R}$ whilst if $\rho(\mu) > \frac{p}{q}$, then $F^q(x) > x+p$ for all $x \in \mathbb{R}$. Thus (see fig. 2.9) either there is a range of parameters such that the graph of $F^q(x) - x$ passes across p, or there is one parameter at which $F^q_\mu(x) - x \equiv p$ for all x. But this latter condition is very unlikely (the qth iterate would be identically linear).

2.4.2 Rotation renormalization and codings

Suppose $f: [0,1] \to [0,1]$ is a rotation-like piecewise-smooth map, i.e., there exists $c \in (0,1)$ such that f is continuous and strictly increasing on $(0,c)$ and on $(c,1)$ with

$$\lim_{x\uparrow c} f(x) = 1, \quad \lim_{x\downarrow c} f(x) = 0, \quad f(0) = f(1).$$

Then we can associate f with a circle homeomorphism with lift F defined by

$$F(x) = \begin{cases} f(x) & \text{if } 0 \le x < c, \\ 1 & \text{if } x = c, \\ f(x) + 1 & \text{if } c < x < 1, \end{cases} \tag{2.48}$$

and $F(x+1) = F(x)+1$. Thus we can talk about the rotation number of f, though F is not necessarily C^2 at integer values, so a little care needs to be taken about bifurcations (this will be considered in the next subsection). The rotation number can also be thought of as

$$\rho(f) = \lim_{n\to\infty} \tfrac{1}{n}\#\{r \mid f^r(x) > c,\ r = 1,2,\dots,n\}$$

since the lift moves solutions into the next interval $(m, m+1)$ if and only if $x > c$.

There is a natural renormalization for circle maps that can help describe the dynamics of examples.

First note that there is a simple trichotomy:

(i) $f(0) = c$; or

(ii) $f(0) > c$; or

(iii) $f(0) < c$.

If $f(0) = c$, then $f((0,c) = (c,1)$ and $f(c,1) = (0,c)$ so $\rho(f) = \frac{1}{2}$.

If $f(0) > c$, then $f((0,c)) \subset (c,1)$ and hence $f^2|(0,c)$ is continuous and monotonic. Either f has a fixed point (and hence rotation number zero or one) or consider the induced map

$$g_L(x) = \begin{cases} f^2(x) & \text{if } 0 \le x < c, \\ f(x) & \text{if } c < x \le f(0). \end{cases} \tag{2.49}$$

This is has two monotonic continuous branches and the image of the left end point, 0, is $g_L(0) = f^2(0)$ whilst the image of the left end-point, $f(0)$, is $g_L(f(0)) = f^2(0)$, so the two end-points map to the same point. Thus after rescaling the interval $[0, f(0)]$ to $[0,1]$ the induced map g_L is again a circle map.

Similarly, if $f(1) = f(0) < c$, then $f((c,1)) \subseteq (0,c)$ and the map

$$g_R(x) = \begin{cases} f(x) & \text{if } f(0) \le x < c, \\ f^2(x) & \text{if } c < x \le 1, \end{cases} \tag{2.50}$$

is (after rescaling) a circle map.

These induced maps provide a way of introducing renormalization ideas to circle maps. If $\rho < \frac{1}{2}$, the coding of orbits for the induced or renormalized map can be used to obtain the coding for the original map by the replacement operation

$$0 \to 0, \quad 1 \to 10. \tag{2.51}$$

In other words, every symbol 1 for the map is followed by a zero. Similarly if $\rho > \frac{1}{2}$, then the replacement operation is

$$0 \to 01, \quad 1 \to 1, \tag{2.52}$$

i.e., every 0 is followed by a 1. This process continues and the symbol sequences obtained in this way have many beautiful properties that have been discovered and rediscovered many times. One particularly nice property is that these sequences are minimax. Let $\Sigma_{p,q}$ with $0 < p < q$ denote all the infinite sequences s of 0s and 1s with period q and which have p 1s in every q symbols, so it has rotation number p/q. Let σ be the standard shift map. Now define

$$s_{p,q} = \min_{s \in \Sigma_{p,q}} \left(\max_{1 \le k < q} \sigma^k s \right). \tag{2.53}$$

Such a sequence is called a minimax sequence of length q.

Then every periodic point with rotation number p/q has a symbolic description that is a shift of the minimax sequence $s_{p,q}$. These sequences are sometimes called rotation compatible sequences, and limits as $p/q \to \omega$ for irrational numbers ω can be taken to define rotation compatible sequences with irrational rotation numbers.

An Example

The golden mean rotation is the rigid rotation map (2.44) with

$$\alpha_* = \tfrac{1}{2}(\sqrt{5} - 1) \approx 0.618.$$

From (2.46) the 'discontinuity' of this map, seen as a map of the interval, is at $c = 1 - \alpha_* < \alpha_*$, so $f(0) > c$ and g_L is the appropriate induced map. Now, without going through the detailed re-scaling to a map of the unit interval, the rotation number of g_L can be read off by noting that geometrically the rotation number *of a map with constant slope* 1 is just the ratio of the height of the image of the left-hand boundary to the length of the interval on which the map is defined, i.e., in the case of g_L it is

$$\frac{g_L(0)}{f(0)} = \frac{2\alpha_* - 1}{\alpha_*} = 2 - \frac{1}{\alpha_*} = 1 - \alpha_*,$$

where we have used the fact that for the golden mean rotation $\alpha_*^2 + \alpha_* - 1 = 0$ to replace $\frac{1}{\alpha_*}$ by $\alpha_* + 1$.

Hence the rescaled induced map g_L has rotation number $1 - \alpha_* < \frac{1}{2}$ and hence it can be renormalized with g_R in (2.50), and if f represents the rescaled version of g_L (the rigid rotation with rotation number $1 - \alpha_*$) the geometric argument shows that the rotation number of the renormalized map is

$$\frac{g_R(f(0)) - f(0)}{1 - f(0)} = \frac{2(1 - \alpha_*) - (1 - \alpha_*)}{1 - (1 - \alpha_*)} = \frac{1 - \alpha_*}{\alpha_*} = \alpha_*.$$

Hence, after two renormalizations the rescaled induced map is again a golden mean rotation. In other words, the golden mean rotation can be renormalized infinitely often using g_L and then g_R repeatedly. This means that from (2.52) and (2.51) it is a fixed point of the replacement operation

$$0 \to 01, \quad 1 \to 101 \tag{2.54}$$

and the full golden mean symbolic sequence can be built up by repeated application of this replacement operation as for period-doubling in Section 2.3.3 yielding

$$0 \to 01 \to 01101 \to 0110110101101 \to \cdots.$$

Shifts of this sequence can also be used to represent the golden mean rotation. To check that this does indeed give a rotation number equal to α_* the same techniques as used in Section 2.3.3 can be used. Let ℓ_n denote the number of 0s in the nth sequence generated by repeated application of (2.54) to 0 and r_n the number of 1s, so the rotation number of the periodic sequence at the nth level is $\rho_n = r_n/(\ell_n + r_n)$. At the next level $r_{n+2} = \ell_n + 2r_n$ as there is one 1 in the replacement symbol for 0 and two 1s in that of 1 in (2.54). The total length of the $(n+1)^{th}$ sequence is $2\ell_n + 3r_n$. Therefore

$$\rho_{n+1} = \frac{r_{n+1}}{ell_{n+1} + r_{n+1}} = \frac{\ell_n + 2r_n}{2\ell_n + 3r_n} = \frac{(\ell_n + r_n) + r_n}{2(\ell_n + r_n) + r_n},$$

and dividing the numerator and denominator by $\ell_n + r_n$,

$$\rho_{n+1} = \frac{1 + \rho_n}{2 + \rho_n}.$$

This has a fixed point ρ_* where $\rho_* = \frac{1+\rho_*}{2+\rho_*}$ or $\rho_*^2 + \rho_* - 1 = 0$, the quadratic equation for α_*.

Note that another symbolic approach is obtained by the observation that after the transformation $x \to 1 - x$, which switches the roles of symbols 0 and 1, a map with rotation number $1 - \alpha_*$ becomes a map with rotation number α_*, and hence we need only renormalize once and then reverse orientation, leading to the replacement operation

$$0 \to 1, \quad 1 \to 01$$

and then

$$1 \to 01 \to 101 \to 01101 \to 10101101 \to 0110110101101 \to \cdots$$

which oscillates between codes tending to the code for α_* and for $1 - \alpha_*$.

2.4.3 Gap maps

Consider maps $f\colon [0,1] \to [0,1]$ such that there exists $c \in (0,1)$ such that f is continuous and strictly increasing on $(0,c)$ and on $(c,1)$, and

$$\lim_{x\uparrow c} f(x) = 1, \quad \lim_{x\downarrow c} f(x) = 1, \quad f(0) < f(1). \tag{2.55}$$

The final condition of (2.55) explains why these maps are called gap maps: there is an interval $(f(0), f(1))$ which has no preimages under f. As with circle maps we can associate f with a lift F as in (2.48), but this time there is a discontinuity at integer values of x. Note that as a map of the interval the discontinuity is at $x = c$, but as a map of the circle it is the gap condition that creates the discontinuity:

$$\lim_{x\uparrow 1} F(x) = 1 + f(1) < 1 + f(0) = \lim_{x\downarrow 1} F(x).$$

However, although the lift of f is discontinuous, the function F is monotonic increasing regardless of the choice made for the value at $x = p$ between the two limiting choices determined by continuity. It is therefore natural to ask whether the results for standard circle maps holds for these discontinuous lifts.

Theorem 2.4.5. *If F is the lift of a gap map, then $\rho(F)$ exists and is independent of both x and the choice of $F(1) \in (1 + f(0), 1 + f(1))$.*

Note that this result is no longer true for maps with gaps and plateaus (intervals on which F is constant), but it remains true if F is strictly increasing and has a countable set of discontinuities.

Definition 2.4.6. (f_μ) is a continuous family of gap maps for $\mu \in (\mu_1, \mu_2) = \mathcal{M}$ if, for all $\mu_0 \in \mathcal{M}$ and all $x \in \mathbb{R}$, $\lim_{\mu\to\mu_0} |F_\mu(x) - F_{\mu_0}(x)| = 0$.

Rhodes–Thompson [83, 84] prove that the bifurcation structure in terms of continuity of rotation numbers and mode-locking is also retained for continuous families of gap maps.

Theorem 2.4.7. *If (f_μ) is a continuous family of gap maps, then $\rho(f_\mu) = \rho(\mu)$ varies continuously. If there exist $\mu_1 < \mu_2$ such that $\rho(\mu_1) < \frac{p}{q} < \rho(\mu_2)$, then typically $\rho(\mu) = \frac{p}{q}$ on an interval of parameter values.*

As before, this implies that if the rotation number varies then non-periodic (irrational rotation number) behaviour is possible though this will be on a Cantor set. These can be very hard to observe numerically, and there was at one stage some confusion as to whether they exist or not.

2.4.4 Overlap maps

An overlap map is a map satisfying the conditions for a gap map but for which the last criterion of (2.55) is replaced by

$$f(0) < f(1). \tag{2.56}$$

Thus rather than having a gap there is a set of points with two preimages under f. These maps can be chaotic and the non-wandering set can be described by kneading theory [73]. In this section we will continue the analogy with circle maps to provide a different view of the effect of overlap. The analogy is with continuous non-invertible circle maps. For these maps the idea of a rotation number is replaced by a rotation interval.

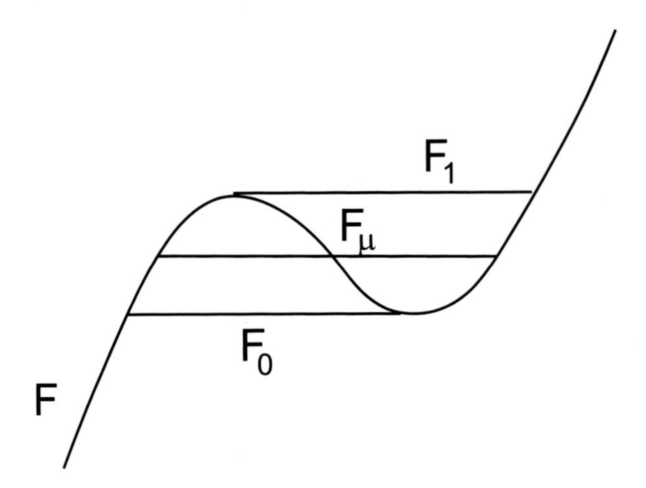

Figure 2.10: Part of the graph of the lift F showing the construction of the maps F_1, F_μ and F_1 which differ from F only on the plateaus.

Definition 2.4.8. The rotation set of a lift F is the set

$$\rho(F) = \{\alpha \mid \rho(x,f) = \alpha \text{ for some } x \in \mathbb{R}\}.$$

The following theorem was proved by Alsedà et al [4] which is analogous to the equivalent result for non-invertible circle maps.

Theorem 2.4.9. *If f is an overlap map with lift F, then $\rho(F)$ is a closed interval (possibly a point).*

Proof (sketch). First note that the lift of f jumps down at integers, for example at $x = 1$ the jump is from $1 + f(1)$ to $1 + f(0)$, so it is no longer strictly increasing and the previous results cannot be used. However, the graph is bounded by two continuous monotonic (but not strictly monotonic) lifts:

$$F_0(x) = \inf_{y>x} F(y), \quad F_1(x) = \sup_{y<x} F(x). \tag{2.57}$$

Clearly (see fig. 2.10) $F_0(x) \leq F_1(x)$ and $F_k(x)$ are increasing and continuous. We will treat the simple case in which there is only one plateau in each period of the lift. Now, F_k can be seen as the inverses of gap maps, and hence (or by direct verification) have well-defined rotation numbers with $\rho(F_0) \leq \rho(F_1)$. Moreover,

$$F_0(x) \leq F(x) \leq F_1(x)$$

implies that for all x such that $\rho(x,F)$ exists, then $\rho(F_0) \leq \rho(x,F) \leq \rho(F_1)$ (indeed we can take limsups and liminfs of $\frac{1}{n}(F^n(x) - x)$ and these will both lie between $\rho(F_0)$ and $\rho(F_1)$). Thus

$$\rho(F) \subseteq [\rho(F_0), \rho(F_1)].$$

To finish we need to show that for all $y \in [\rho(F_0), \rho(F_1)]$ there exists x such that $\rho(x, F) = y$. We begin by interpolating between F_0 and F_1 creating a continuous family F_μ, $0 \leq \mu \leq 1$ of monotonic circle maps as shown in fig. 2.10. Each of these has a unique rotation number $\rho_\mu \in [\rho(F_0), \rho(F_1)]$ and ρ_μ varies continuously with μ by Theorem 2.4.7. Thus for every $r \in [\rho(F_0), \rho(F_1)]$ there exists $\mu \in [0,1]$ such that $\rho_\mu = r$. To complete the proof we will show that for each monotonic circle map f_μ with lift F_μ and plateau with open arc P there exists $x \in \mathbb{T}$ such that $f_\mu^n(x) \notin P$ for all $n \geq 0$ and hence, since $f(x) = f_\mu(x)$ if $x \notin P$, then the orbit of x under f_μ is the orbit of x under f and since $\rho(F_\mu) = \rho_\mu$ exists and is independent of x, $\rho(F, x) = \rho_\mu$.

Let

$$\Gamma_n = \{x \in \mathbb{T} \mid f^k(x) \notin P,\ k = 0, 1, \ldots, n\}.$$

Then since P is open and f_μ and f_μ^{-1} are continuous on $\mathbb{T} \backslash P$, Γ_n is closed and $\Gamma_{n+1} \subseteq \Gamma_n$, hence provided $\Gamma_n \neq \emptyset$ for some n, the limit $\cap \Gamma_n$ is closed and non-empty. Points in this countable intersection have precisely the required property.

So suppose that there exists $m > 0$ such that $\Gamma_m = \emptyset$, i.e., for all $x \in \mathbb{T}$ there exists $k \leq m$ such that $f^k(x) \in P$. Now, $f_\mu(P) = y$ is a point, and hence $\cup_{k \geq 0} f_\mu^k(P) = P \cup C$, where C is a countable set of points, and in particular $P \cup C \neq \mathbb{T}$. But by assumption, for all $x \in \mathbb{T}$, $f_\mu^m(x) \in P \cup C$, i.e., $f_\mu^m(\mathbb{T}) \subseteq P \cup C$. But f_μ is a surjection, so $f_\mu^m(\mathbb{T}) = \mathbb{T}$, hence we have a contradiction. □

The bifurcation structure of these maps was investigated by Gambaudo et al [34] in the 1980s for the case where the slope tends to zero at the discontinuity.

2.5 Gluing bifurcations

Gluing bifurcations describe the dynamics of piecewise monotonic maps near codimension two points for maps that are locally contracting. There are three cases determined by the orientation of each continuous branch of the map. These codimension two bifurcations have been described by various authors, e.g., [38, 53], in recent years, but as a historical curiosity we will follow the account of Glendinning [40] from 1985. This was work done with Gambaudo–Tresser intended to form part of the sequel to [32], but which was never completed. The analysis was in the context of homoclinic bifurcations related to the Lorenz semi-flows of Section 2.1.2, and was done in a hurry to meet Fellowship deadlines so it is not as polished as I would like. However, it does show how much was understood in the mid 1980s. See [53] for a more complete description.

2.5.1 The three cases

Thus the remainder of this section is taken verbatim from [40]. Where there is reference to work elsewhere in the dissertation, or where the context may be unclear I have added commentary in italics inside square brackets [*thus*].

START of excerpt from [40]

In the general case we have two parameters which, as usual, can be thought of as parameterising the x-coordinate of the first intersection of the two branches of the unstable manifold of the stationary point with a surface inside a small neighbourhood of the stationary point. recall that the one dimensional map used to model the flow is a piecewise monotonic function with a single discontinuity (at $x = 0$):

$$x' = \begin{cases} -\mu + ax^\delta, & x > 0, \\ \nu - b(-x)^\delta, & x < 0. \end{cases} \tag{2.58}$$

[*The assumption that $\delta > 1$ was part of the chapter introduction; all the results below depend upon is local monotonicity and contraction.*] Note that the signs of μ and ν have been changed so that the interesting behaviour arises when the parameters are positive. Once again there are four cases depending on the signs of the constants a and b, i.e., whether the global reinjections are orientable (positive signs) or non-orientable (negative signs). Regardless of the signs of a and b we can see from the graph of the map that $\mu < 0$, $\nu < 0$ there is a pair of stable fixed points of the map and so, at least locally, these are the only periodic orbits of the map. The remainder of the parameter space varies according to the signs of a and b so the various cases will be treated separately.

(a) The Orientable case: $a > 0$, $b > 0$

The curve in parameter space given by $\mu = 0$ (resp. $\nu = 0$) corresponds to a line of homoclinic orbits [*i.e., border bifurcations as in Section* 2.3.5] involving the positive (resp. negative) branch of the unstable manifold of the origin. Hence (cf. Section 2.1) we know that on crossing this curve a periodic orbit is generated. From the graph of the model map it is clear that this periodic orbit corresponds to the fixed point in $x > 0$ (resp. $x < 0$) which exists for $\mu < 0$ (resp. $\nu < 0$). Since the slope of the map is always positive and less than one any orbit that enters $x > 0$ when $\mu < 0$ (resp. $x > 0$ when $\nu < 0$) tends directly to the fixed point. Using the standard coding of orbits, 0 for points in $x < 0$ and 1 for points in $x > 0$, these facts imply that the only periodic orbits are

0 and 1	if $\mu < 0$ and $\nu < 0$,
0	if $\mu > 0$ and $\nu < 0$,
1	if $\mu < 0$ and $\nu > 0$.

If μ and ν are both positive, then the situation is considerably more complicated. We shall prove the following theorem.

Theorem 2.5.1 (Theorem 3.2). *For $\mu > 0$ and $\nu > 0$ the periodic orbits of* (2.58) *have the following properties:*

(i) *There is at most one periodic orbit.*

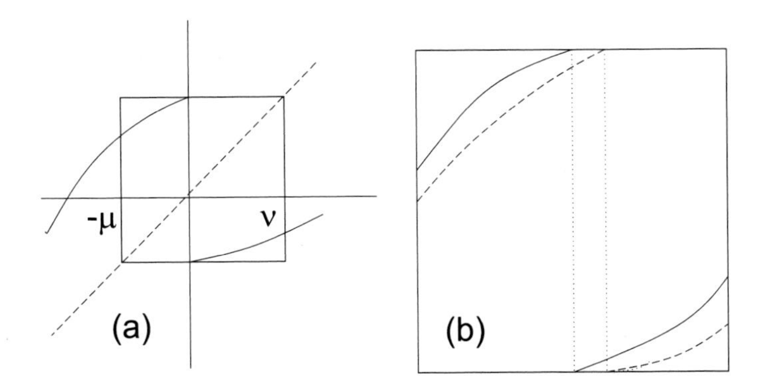

Figure 2.11: ([40, Fig. 65].) (a) The return map with $\mu, \nu > 0$ showing the region $[-\mu, \nu]$ in which the orbits of 0^+ and 0^- remain. (b) The map $g_{\mu,\nu}$ for two values of the parameter ν. The associated lift is monotonic.

(ii) *Periodic orbits have codes which are rotation compatible, i.e., minimax.*

(iii) *The rotation number of periodic orbits varies monotonically with one parameter when the other is held fixed.*

Statements (i) and (ii) are a simple consequence of the theorems of Section 3.1 [*those of Section* 2.4 *above*]. The important new part is (iii). This statement implies that there are parameter values in a neighbourhood of $(0, 0)$ for which the system has a periodic orbit with *any* given minimax code, and also that there are parameter values at which the rotation number of a periodic orbit is irrational, i.e., there are aperiodic orbits which are stable (and so not chaotic). [*That could have been phrased better!*] From the geometry of the flow it is clear that the orbits lie on a torus with a hole. This is precisely the property of Cherry flows which have been studied by many pure mathematicians (e.g., Palis and de Melo, 1984 [82]). It is curious that these apparently abstract flows arise naturally near pairs of homoclinic orbits.

To prove statement (iii) we begin by rescaling the map so that the important dynamics (and in particular the orbits of 0^+ and 0^- [*the limits as* 0 *is approached from above or below respectively*]) is contained in the interval $[-1, 1]$. The iterates of 0^+ and 0^- remain in the interval $[-\mu, \nu]$ (see Fig. 65 [*i.e., Figure* 2.11 *here*] so we look for a change of coordinates of the form $z = p + qx$ such that

$$-1 = p - q\mu$$

and

$$1 = p + q\nu$$

so that when $x = -\mu$, $z = -1$ and when $x = \nu$, $z = 1$. This gives

$$p = (\mu - \nu)/(\mu + \nu),$$
$$q = 2/(\mu + \nu),$$

i.e.,

$$z = \{\mu - \nu + 2x\}/(\mu + \nu). \tag{2.59}$$

In the new coordinates we have the map $g_{\mu,\nu}: [-1,1] \to [-1,1]$ given by

$$g_{\mu,\nu} = \begin{cases} -1 + a[(z - \frac{\mu-\nu}{\mu+\nu})/2]^\delta & \text{if } \frac{\mu-\nu}{\mu+\nu} < z < 1, \\ = 1 - b[(z - \frac{\mu-\nu}{\mu+\nu})/2]^\delta & \text{if } -1 < z < \frac{\mu-\nu}{\mu+\nu}. \end{cases} \tag{2.60}$$

This new map has all the properties of $f_{\mu,\nu}$ and in particular orbits have the same rotation number. Note that $g_{\mu,\nu}$ is piecewise increasing with a single discontinuity and so it can be viewed as a discontinuous map of the circle to itself. Hence we can associate a lift $G_{\mu,\nu}: \mathbb{R} \to \mathbb{R}$ with $g_{\mu.\nu}$ and so define a rotation number in the usual way. Viewing $g_{\mu,\nu}$ as an application [*map in French*] of the circle we have, from Theorem 1 of Gambaudo and Tresser (1985) [36].

- to all $x \in [-1,1]$ there is a unique rotation number $\rho_{\mu,\nu}$ (this follows from the piecewise monotonicity of the mapping);
- for all $\rho_{\mu,\nu}$ there is a rotation compatible orbit with that rotation number.

Given the uniqueness of the rotation number for given values of the parameters and the existence of a rotation compatible orbit with this rotation number, we obtain (i) and (ii) of the theorem.

Gambaudo and Tresser (1985) [36] also show that if the lift of a map depending on a single parameter is increasing with the parameter, then the rotation number is increasing and continuous with the parameter [*for continuous families as shown described in Section* 2.4; *obvious for this family* (2.58) *but needs stating for the more general case*]. A direct application of this result gives part (iii) of the theorem noting that $G_{\mu,\nu}$ is increasing with ν (Fig. 65) [*Figure* 2.11 *here*]. □

Outside the region of validity of this local analysis, outside some neighbourhood of the codimension two homoclinic bifurcation, the appearance of chaotic behaviour can also be studied in a similar way. In the next section we shall discuss the appearance of chaos and look at a simple example. However, first we shall complete the local analysis for the remaining two cases.

(b) The Semi-orientable case: $a > 0$, $b < 0$

Here the right-hand reinjection ($x > 0$) is orientable whilst the other is non-orientable. Using the techniques above we can show from the one-dimensional map that

- in $\mu < 0$, $\nu < 0$, the only periodic orbits have codes 1 and 0;
- in $\mu < 0$, $\nu > 0$, the only periodic orbit has code 1;
- in $\mu > 0$, $\nu < 0$, the periodic orbit with code 0 exists throughout the quadrant, and in $\nu > b\mu^\delta$ there is also a periodic orbit with code 10.

The final quadrant, with both μ and ν positive is more complicated. We shall prove the following theorem:

Theorem 2.5.2 (Theorem 3.3). *For $a > 0$, $b < 0$ in (2.58) and $\mu, \nu > 0$, there is a neighbourhood of $(\mu, \nu) = (0, 0)$ in which the only periodic orbits are those with codes 1^n0, $n \geq 1$ and furthermore, regions of parameter space in which orbits with codes 1^n0 and $1^{n+1}0$ coexist, $n \geq 1$.*

[*This statement is seriously ungrammatical: as the proof below shows it is intended that there are regions with just code 1^n0 and regions with the stated coexistence.*]

First note that the periodic orbits must have all their points in $[-\mu, \nu - b\mu^\delta]$ and that the map is decreasing and positive in $x < 0$, and increasing in $x > 0$. Let $N \geq 2$ be the first time that $f^N(x) < 0$ for some $x \in [-\mu, 0)$ and note that parameter values can be found such that any given value of N (≥ 2) can be realised, with $N = 2$ for μ/ν large and $N \to \infty$ as $\mu/\nu \to 0$. Now consider the map

$$h(x) = \begin{cases} f(x) & \text{if } x > 0, \\ f^N(x) & \text{if } x < 0. \end{cases} \tag{2.61}$$

Since $f'(x) > 0$ in $x > 0$ and $f(x) > 0$ for $x < 0$ we have $f^N(0^-) < f^N(x)$ for x in $[-\mu, 0)$ and, by the definition of N, $f^N(0^-) < 0$. Hence $h(x)$ looks like $f(x)$ (upside down) in the region of parameter space with $\mu > 0$ and $\nu < 0$. The remarks made above for this quadrant of parameter space hold: there is a periodic orbit with code 0 and, in some cases, a periodic orbit with code 10, for h. In terms of f this translates into the existence of a periodic orbit with code 01^{N-1} and, in some cases, a periodic orbit with code 01^N coexists with this first orbit. It should be clear that on a line in parameter space with ν constant both possibilities must be realized, hence the theorem.

Now, the homoclinic orbit associated with a periodic orbit of code 10 occurs when $f(\nu) = -\mu + a\nu^\delta = 0$, i.e., $\mu = a\nu^\delta$. This gives the bifurcation diagram in fig. 66 [fig. 2.12 *here, remark about shading deleted*].

(c) The Non-orientable case: $a < 0$, $b < 0$

When both the global reinjections are non-orientable, the dynamics of the local map is relatively simple and we obtain essentially the same diagram as fig. 53 [fig. 2.13 *here*] for the orientable figure eight. It follows directly from the one-dimensional map (2.58) that

- if $\mu < 0$ and $\nu < 0$, the only periodic orbits have codes 1 and 0;

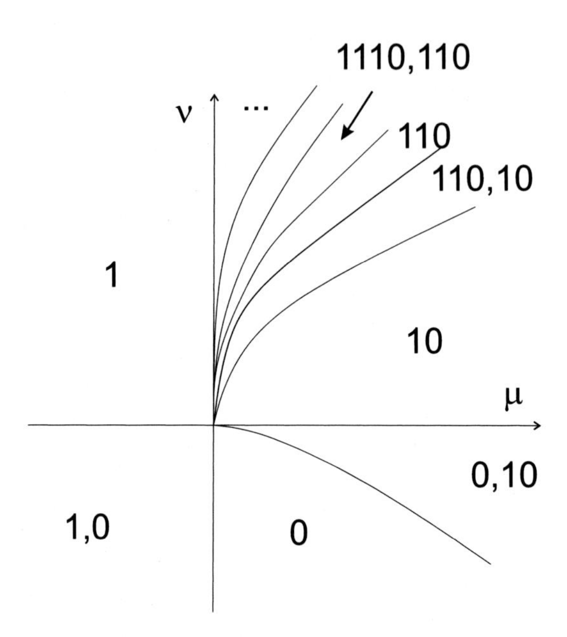

Figure 2.12: ([40, fig. 66], with minor modification.) (μ, ν) parameter for the semi-orientable case.

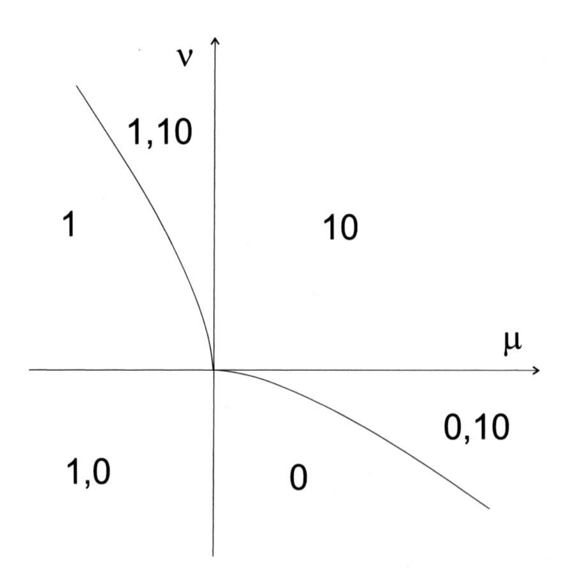

Figure 2.13: ([40, fig. 53].) (μ, ν) parameter space showing the homoclinic curves and the codes of periodic orbits for the orientable case [*of the figure eight configuration*].

- if $\mu > 0$ and $\nu < 0$, the periodic orbit with code 0 exists throughout the region and a periodic orbit with code 01 coexists with it if $-\mu > a\nu^\delta$;
- if $\mu < 0$ and $\nu > 0$, the periodic orbit with code 1 exists throughout the region and a periodic orbit with code 10 coexists with it if $-\nu > b\mu^\delta$;
- if $\mu > 0$ and $\nu > 0$, the periodic orbit with code 01 exists throughout the region and is the only periodic orbit of the local analysis.

This completes the local bifurcation pictures for the butterfly configuration of homclinic orbits.

END of excerpt from [40]

The extract above is an early draft and could obviously be improved (it was written to a deadline). But it does indicate the results we understood at that time. Further details including diagrams for the orientable case and links with differential equations can be found in [33] and the cases were described again in [31]. An excellent modern account can be found in [53].

2.5.2 Anharmonic cascades

We have encountered a number of situations where there are infinite sequences of bifurcations accumulating at some value of the parameter:

(i) period-doubling cascades in smooth maps (a transition to chaos);

(ii) period-adding cascades in the square root map (noting that these refer to stable periodic orbits only);

(iii) anomalous doubling in an unbounded map (noting that this reduced to a version of period-doubling in an induced map);

(iv) continuous change of rotation number in gap maps (no chaos);

(v) period-adding in gluing bifurcations (again, with no chaos).

2.6 Piecewise-smooth maps of the plane

The results in previous sections rely heavily on the order property of the real line, and maps in the plane are much harder to analyze. In many ways this section is a list of results and techniques without a strong over-arching theory underpinning it. We will start with some examples and phenomenology.

2.6.1 The Lozi map

The Lozi map is a natural extension of tent maps to the plane. If the family of tent maps is written in the form

$$x_{n+1} = 1 - a|x_n|, \quad a \in (1,2], \tag{2.62}$$

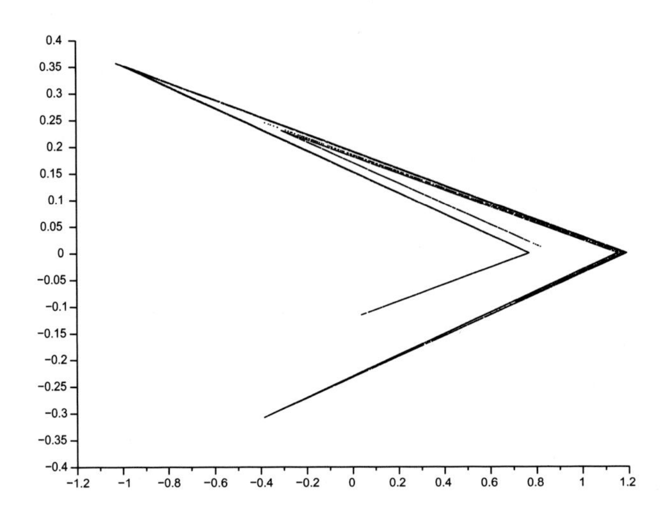

Figure 2.14: Numerically computed attractor of the Lozi map with $a = 1.7$ and $b = 0.3$.

then the Lozi map is the map of the plane defined by

$$\begin{aligned} x_{n+1} &= 1 - a|x_n| + y_n \\ y_{n+1} &= bx_n. \end{aligned} \tag{2.63}$$

If $b \to 0$, then $|y_n| \to 0$ and so the x evolution is the tent map (2.62). If $b \neq 0$ is small, then the attractor looks like a set of folded lines as shown in fig. 2.14. The Lozi map has uniform expansion and contraction properties, and this makes it a good example to test our ability to prove the existence of strange attractors.

2.6.2 The border collision normal form

The border collision normal form (BCNF) is a generalization of the Lozi map that describes the local behaviour of piecewise-smooth maps for which a fixed point of one of the smooth systems defining the map hits a boundary on which that system is defined as parameters are varied. It is usually written as

$$\begin{pmatrix} x_{n+1} \\ y_{n+1} \end{pmatrix} = \begin{cases} f_0(x_n, y_n) & \text{if } x < 0, \\ f_1(x_n, y_n) & \text{if } x_n > 0, \end{cases} \tag{2.64}$$

with

$$f_k(x,y) = \begin{pmatrix} T_k & 1 \\ -D_k & 0 \end{pmatrix} \begin{pmatrix} x_n \\ y_n \end{pmatrix} + \begin{pmatrix} \mu \\ 0 \end{pmatrix}, \quad k = 0, 1. \tag{2.65}$$

The constants T_k and D_k are the trace and determinant of the matrix and μ is the bifurcation parameter. Note that by scaling x and y the parameter μ can be taken in the set $\{-1, 0, +1\}$, so the idea that μ varies continuously is unnecessary.

The BCNF is continuous, but (assuming that $T_0 \neq T_1$ or $D_0 \neq D_1$) the Jacobians are different. Even so, the number of different phenomena that can be observed is huge. In [9] eight different transitions involving observed attractors are mentioned explicitly:

$\mu < 0$	to	$\mu > 0$
no attractor		fixed point
no attractor		chaos
fixed point		chaos
fixed point		fixed point
fixed point and period 3		fixed point and period 4
fixed point		period 2
fixed point and period 11		period 2
fixed point		period 5 and chaos

With all this complexity it can be very difficult to decide what to analyze mathematically. Rather than attempt a complete classification we will give examples of the sort of thing that can be done in each of the three cases: periodic, one-dimensional and two-dimensional attractors in the next section.

2.6.3 Fixed points and the stability triangle

Fixed points, periodic orbits and linear stability of smooth maps can be treated in the same way as the one-dimensional case, but the derivative is replaced by the Jacobian matirx. Given

$$\boldsymbol{x}_{n+1} = f(\boldsymbol{x}_n)$$

with $f: \mathbb{R}^2 \to \mathbb{R}^2$ a fixed point satisfies $x = f(x)$ and assuming this is not on the boundary stability is determined (via a standard small perturbation argument) by the eigenvalues of the Jacobian matrix

$$Df(x) = \begin{pmatrix} \frac{\partial f_1}{\partial x} & \frac{\partial f_1}{\partial y} \\ \frac{\partial f_2}{\partial x} & \frac{\partial f_2}{\partial y} \end{pmatrix}$$

with partial derivatives evaluated at the fixed point.

Eigenvalues of the Jacobian matrix satisfy the characteristic equation $s^2 - Ts + D = 0$, where T is the trace of the Jacobian and D the determinant. The fixed point is stable if the eigenvalues

$$s_{\pm} = \tfrac{1}{2}(T \pm \sqrt{T^2 - 4D})$$

lie inside the unit circle. The boundary of the region of parameters in (T, D) space for which the fixed point is stable has three components:

(i) *Complex conjugate eigenvalues.* If $T^2 < 4D$, then $s_\pm$ are complex conjugates and $|s_\pm|^2 = s_+ s_- = D$ as $s^2 - Ts + D = 0$. Thus our first stability criterion is

$$T^2 < 4D, \quad D < 1. \tag{2.66}$$

(ii) *Real eigenvalues: positive trace.* If $T^2 > 4D$, then $s_\pm$ are real and if $T > 0$, then $|s_+| > |s_-|$ and so both are less than 1 if

$$s_+ = \tfrac{1}{2}(T + \sqrt{T^2 - 4D}) < 1,$$

i.e., $\sqrt{T^2 - 4D} < 2 - T$ so $T < 2$ and $T^2 - 4D < 4 - 4T + T^2$, i.e., $T < 1 + D$. Thus our second stability criterion is

$$0 < T < 2, \qquad T^2 > 4D, \qquad T < 1 + D. \tag{2.67}$$

(iii) *Real eigenvalues: negative trace.* If $T^2 > 4D$ then $s_\pm$ are real and if $T < 0$ the $|s_-| > |s_+|$ and so both are less than one if

$$s_- = \tfrac{1}{2}(T - \sqrt{T^2 - 4D}) > -1,$$

i.e., $\sqrt{T^2 - 4D} < 2 + T$ so $T > -2$ and $T^2 - 4D < 4 + 4T + T^2$, i.e. $-T < 1 + D$. Thus the third stability criterion is

$$-2 < T < 0, \quad T^2 > 4D, \quad -T < 1 + D. \tag{2.68}$$

Putting the three criteria (2.66)-(2.68) together we find

$$T^2 < 4D, \quad D < 1,$$
$$T^2 > 4D, \quad 0 < |T| < 2, \quad |T| < 1 + D,$$

which can be written more succinctly as

$$D < 1, \quad |T| < 1 + D. \tag{2.69}$$

This describes three straight lines in the (T, D) bifurcation plane which bound a triangle inside of which the fixed point is linearly stable.

More generally there are five generic linear types of fixed points in the plane

(i) stable/unstable foci ($s\pm$ complex conjugate pair);

(ii) stable/unstable node ($s_\pm$ real distinct, $|s_\pm| < 1$);

(iii) saddle.

Phase portraits are very similar to the analogous continuous time stationary points, though solutions 'jump' along the continuous curves in discrete time. Negative real values of s allow solutions to jump between the integral curves of the continuous time analogues and it changes the sign of the variable at each iteration.

As in the one-dimensional case periodic points can be thought of as are fixed points of f^p, and the linear stability of a periodic orbit is analyzed by looking at the eigenvalues of Df^p.

Fixed points of continuous piecewise-smooth systems

Choose co-ordinates with the switching surface at $x = 0$ so

$$\boldsymbol{x}_{n+1} = f(\boldsymbol{x}_n) = \begin{cases} f_0(x,y) & \text{if } x < 0, \\ f_1(x,y) & \text{if } x \geq 0. \end{cases}$$

Then continuity implies that $f_0(0,y) = f_1(0,y)$. (This mild form of piecewise-smooth system is already complicated enough without adding discontinuity.)

Fixed points with $x \neq 0$ have the same local structure as just described, and fixed points on the switching surface are typically of codimension one and so an exhaustive classification is not warranted.

However, as soon as one starts to consider one-parameter families of piecewise-smooth maps, codimension one phenomena become typical and although a detailed knowledge of what happens at this parameter may be unimportant, the effect of the codimension one situation locally *is* important — this is precisely the point of bifurcation theory. Thus, at least from the theoretical viewpoint learned in smooth bifurcation, an analysis of the local effect of having a codimension one system which has a fixed point on the switching surface should be considered. Such codimension one systems give rise to *border collision bifurcations*, the two-dimensional equivalent of the bifurcations described in Section 2.3.5 for one-dimensional systems.

2.6.4 Border collision bifurcations

Consider one-parameter families of continuous piecewise-smooth maps:

$$\boldsymbol{x}_{n+1} = f(\boldsymbol{x}_n, \mu) = \begin{cases} f_0(x,y,\mu) & \text{if } x < 0, \\ f_1(x,y,\mu) & \text{if } x \geq 0, \end{cases} \tag{2.70}$$

so continuity implies that

$$f_0(0,y,\mu) = f_1(0,y,\mu). \tag{2.71}$$

Assume that there is a fixed point on the switching surface $x = 0$ if $\mu = 0$,

$$f_0(0,0,0) = f_1(0,0,0) = 0, \tag{2.72}$$

What happens locally?

Nusse–Yorke [81] show that the local behaviour can be described (to lowest order terms) by the *border collision normal form*

$$\begin{pmatrix} x_{n+1} \\ y_{n+1} \end{pmatrix} = \begin{cases} f_0(x_n,y_n) & \text{if } x < 0, \\ f_1(x_n,y_n) & \text{if } x_n > 0, \end{cases}$$

with

$$f_k(x,y) = \begin{pmatrix} T_k & 1 \\ -D_k & 0 \end{pmatrix} \begin{pmatrix} x \\ y \end{pmatrix} + \begin{pmatrix} \mu \\ 0 \end{pmatrix}, \quad k = 0,1.$$

This is such an important set of equations that we will show how it is derived from the general forms (2.70), (2.71) and (2.72); this is also a good exercise which demonstrates how coordinate transformations are constrained by keeping the switching surface in a simple (i.e., unchanged in this case) form.

By Taylor expansion

$$f_k(x,y) = \begin{pmatrix} a_k & s \\ b_k & t \end{pmatrix} \begin{pmatrix} x \\ y \end{pmatrix} + \mu \begin{pmatrix} u \\ v \end{pmatrix}, \quad k = 0,1.$$

Where we have ignored quadratic terms and higher, and t, s, u, v are independent of k by continuity. Then the general coordinate transform which keeps $x = 0$ invariant is:

$$Y = \alpha x + y, \quad X = \beta x$$

with $\beta \neq 0$ and the coefficient of y non-zero (so that the coordinate transformation is invertible, i.e., the coordinates are for independent) and thus can be chosen to be unity by scale invariance.

Now we just go through the calculation!

$$f_k(x,y) = \begin{pmatrix} T_k & 1 \\ -D_k & 0 \end{pmatrix} \begin{pmatrix} x \\ y \end{pmatrix} + \begin{pmatrix} \mu \\ 0 \end{pmatrix}, \quad k = 0,1.$$

Note that

(i) by changing scale by a factor of $|\mu|$: without loss of generality $\mu \in \{-1,0,1\}$;

(ii) if $D_1 > 0$, then right half-plane mapped to lower half-plane; upper half-plane if $D_1 < 0$;

(iii) if $D_0 > 0$, then left half-plane mapped to upper half-plane; lower half-plane if $D_0 < 0$;

(iv) BCNF a homoemorphism if $D_0 D_1 > 0$; non-invertible if $D_0 D_1 < 0$.

2.7 Periodic orbits and resonance

Suppose that one fixed point of the border collision normal form exists and the Jacobian has complex eigenvalues. In that case the motion on one side of the switching surface is like a rotation, with orbits spiralling in or out of the fixed point. It is then natural to look for periodic orbits that can be described by sequences of 1s and 0s (reflecting iterates in $x > 0$ and $x < 0$ respectively) that come from the order of rotations described in Section 2.4.2. It turns out that this can be carried out exactly, giving equations determining when these orbits exist.

2.7.1 Fixed points and period two

Before considering the more complicated orbits it clearly makes sense to look at simplest orbits: fixed points and points of period two. For non-degenerate systems solutions to linear equations are unique, and so the only period two orbits that we will look for ar those with one point in $x < 0$ and one point in $x > 0$.

A fixed point exists in $x > 0$ if there is a solution to $f_1(x,y) = (x,y)$ with $x > 0$, i.e., if

$$T_1 x + y + \mu = x, \quad -D_1 x = y, \quad x > 0.$$

Solving these simple linear equations gives

$$x = \frac{\mu}{1 + D_1 - T_1}, \quad y = -\frac{D_1 \mu}{1 + D_1 - T_1} \tag{2.73}$$

and so provided $1 + D_1 - T_1 \neq 0$ there is a fixed point for an appropriate sign of μ: $\mu > 0$ if $1 + D_1 - T_1 > 0$ and $\mu < 0$ if $1 + D_1 - T_1 < 0$.

A precisely analogous manipulation shows that there is a fixed point in $x < 0$ provided $\mu > 0$ if $1 + D_0 - T_0 < 0$ and $\mu < 0$ if $1 + D_0 - T_0 > 0$. Moreover, a fixed point is stable if the modulus of every eigenvalue of the Jacobian is less than one. This translates to the conditions $|D_0| < 1$ and $|T_0| < 1 + D_0$ in $x < 0$. This condition can also be written as $0 < 1 + D_0 - |T_0|$. Precisely analogous conditions hold in $x > 0$.

Putting the two branches of solutions together we see that if

$$(1 + D_0 - T_0)(1 + D_1 - T_1) > 0, \tag{2.74}$$

then the two fixed points exist for opposite signs of μ whilst if

$$(1 + D_0 - T_0)(1 + D_1 - T_1) < 0, \tag{2.75}$$

then the two fixed points exist for the same sign of μ (rather like a smooth saddle-node bifurcation).

Except in the degenerate case that a Jacobian has an eigenvalue of -1, in which case there can be a degenerate line of orbits of period two, an orbit of period two has one point on each side of $x = 0$. The equations are a little more messy, but still linear. Going through the detailed calculation period two points are at (x_0, y_0) with $x_0 < 0$ and (x_1, y_1) with $x_1 > 0$ and

$$\begin{aligned} x_0 &= \mu + y_1 + T_1 x_1, & y_0 &= -D_1 x_1, \\ x_1 &= \mu + y_0 + T_0 x_0, & y_1 &= -D_0 x_0, \end{aligned} \tag{2.76}$$

which imply

$$(x_k, y_k) = \left(\frac{1 + T_{1-k} + D_{1-k}}{(1 + D_0)(1 + D_1) - T_0 T_1} \mu, -D_{1-k} x_{1-k} \right), \quad k = 0, 1. \tag{2.77}$$

These lie on the 'correct' sides of the y-axis for one sign of μ provided

$$(1 + T_0 + D_0)(1 + T_1 + D_1) < 0, \tag{2.78}$$

and if this inequality does not hold, then there are no non-degenerate points of period two.

Stability is determined by the trace and determinant of the product of the linear parts of the BCNF:

$$\begin{pmatrix} T_0 & 1 \\ -D_0 & 0 \end{pmatrix}\begin{pmatrix} T_1 & 1 \\ -D_1 & 0 \end{pmatrix} = \begin{pmatrix} T_0T_1 - D_1 & T_1 \\ -D_0T_1 & -D_0 \end{pmatrix}.$$

and the period two orbit is stable if the modulus of the trace and the modulus of the determinant satisfy equivalent conditions as for the fixed points; i.e., it is stable if

$$|D_0 + D_1 - T_0T_1| < 1 + D_0D_1, \quad |D_0D_1| < 1. \tag{2.79}$$

So much for the equations — but what combinations of fixed points and periodic orbits can be involved in bifurcations? This is not obvious from the equations. We leave this question as an exercise for the moment, and will return to it in Section 2.10.2.

2.7.2 Periodic orbits

Although a great deal was known about periodic orbits and the regions of parameter space for which they exist (and may coexist) from the works of Gardini and others [37, 74], the more recent approach of Simpson [85] and Simpson–Meiss [88] makes a systematic approach possible.

Let $s_1, \ldots, s_n$ be a sequence of 0s and 1s, and suppose we wish to look for a periodic orbit of period n such that the k-th point of the periodic orbit lies in $x < 0$ if $s_k = 0$ and in $x > 0$ if $s_k = 1$. To find such an orbit it is necessary to solve the fixed point equation for the n-th iterate of the map, taking into account the required sequence (s_k), and then to determine whether the fixed point (ia periodic orbit of f) is *real*, i.e., its orbit passes through the regions $x \le 0$ and $x \ge 0$ in the prescribed order, or *virtual*, otherwise, in which case the solution does not correspond to an orbit of the BCNF.

At each iteration $f(\boldsymbol{x}) = A_{s_k}\boldsymbol{x} + \mu\boldsymbol{e}$ and so by induction

$$f^n(\boldsymbol{x}) = M_s\boldsymbol{x} + \mu P_s\boldsymbol{e},$$

where

$$M_s = A_{s_n}\cdots A_{s_1}, \quad P_s = I + A_{s_n} + A_{s_n}A_{s_{n-1}} + \cdots + A_{s_n}\cdots A_{s_2}.$$

The point calculated on the orbit of period n in the half plane determined by s_1 is a solution of the fixed point equation $\boldsymbol{x}_1 = f^n(\boldsymbol{x}_1)$, i.e.,

$$\boldsymbol{x}_1 = \mu(I - M_s)^{-1}P_s\boldsymbol{e}. \tag{2.80}$$

Of course, this exists and is unique if $I - M_s$ is non-singular, or eqivalently if $\det(I - M_s) \neq 0$.

The same process can be repeated for each point on the orbit: the image of $\boldsymbol{x}_1$ is $\boldsymbol{x}_2$ which satisfies a similar equation but with s replaced by $s_1 s_n \cdots s_2$. Define the shift σ on these periodic sequences so that

$$\sigma(s_n \cdots s_2 s_1) = s_1 s_n \cdots s_2,$$

then the n points on the orbit of period n corresponding to s are

$$\boldsymbol{x}_{k+1} = \mu(I - M_{\sigma^k s})^{-1} P_{\sigma^k s} \boldsymbol{e}, \qquad k = 0, 1, \ldots, n-1. \tag{2.81}$$

Simpson and Meiss [89] show that the x coordinate of (2.81) can be written as

$$x_{k+1} = \mu \frac{\det P_{\sigma^k s}}{\det(I - M_s)}, \qquad k = 0, 1, \ldots, n-1, \tag{2.82}$$

where we have used the fact that $\det(I - M_{\sigma^k s})$ is independent of k (to see this simply note that $A_{s_1}(I - M_s)A_{s_1}^{-1} = I - M_{\sigma s}$). The remainder of the derivation is far from trivial and details can be found in [89].

So far, so much manipulation. But is this solution real or virtual? The answer is very similar to that in the case of the orbit of period two in Section 2.7.1.

Lemma 2.7.1. *Fix $s = s_1 \cdots s_n \in \{0,1\}^n$ and suppose that $\det(I - M_s) \neq 0$ and $\det P_{\sigma^k s} \neq 0$, $k = 0, 1, \ldots, n-1$. If there exists $g \in \{-1, 1\}$ such that*

$$\begin{aligned} \text{sign}\,(\det P_{\sigma^{k-1} s}) &= -g \quad \text{if } s_k = 0, \\ \text{sign}\,(\det P_{\sigma^{k-1} s}) &= g \quad \text{if } s_k = 1, \end{aligned} \tag{2.83}$$

then the periodic orbit corresponding to s exists for $\mu > 0$ if $g \det(I - M_s) > 0$ and for $\mu < 0$ if $g \det(I - M_s) < 0$.

The proof is straightforward from the definitions and (2.82). Note that at this stage we have not used the assumption that the map is two-dimensional.

2.7.3 Resonance tongues and pinching

Lemma 2.7.1 and (2.82) show that the ways by which periodic orbits can be created or destroyed as parameters vary must involve one or other of P_s or $I - M_s$ becoming singular as the parameters are varied.

One fairly general case of this has some interesting and immediately recognisable features. Simpson and Meiss [89] describe regions in the parameter space of the BCNF in which the map has periodic orbits of particular types. These regions have an interesting structure: the resonant tongues shrink to zero width at some places, creating a picture a bit like a string of sausages. The parameters are chosen so that

$$T_L = 2r_L \cos(2\pi\omega), \quad D_L = r_L^2, \quad T_R = \frac{2}{s_R} \cos(2\pi\omega), \quad D_R = \frac{1}{s_R^2}$$

and $r_L = 0.2$, $\mu = 1$. The bifurcation diagrams of [89] describe the periodic orbits in the (ω, s_R) parameter space and the pinched structure of the tongues can be understood using the methods of the previous section.

The analysis of these bifurcations involves two ingredients. First, the rotation order of the periodic points implies that the points on the periodic orbit can be arranged on a circle (with no self-intersections) so that the order on the circle is $\boldsymbol{x}_1, \dots, \boldsymbol{x}_n$ and the effect of the map f is

$$f(\boldsymbol{x}_k) = \boldsymbol{x}_{k+m} \tag{2.84}$$

where the index $k + m$ is interpreted modulo n with the convention that $0 \equiv n$. However, this order does not indicate where the switching surface lies, so the second ingredient specifies the position of the switching surface with respect to the periodic points. Assume that the switching surface separates the periodic points into two consecutive sets of points on the circle with ℓ points in $x < 0$ and $n - \ell$ in $x > 0$. The orbit is therefore specified by three positive integers: n, m and ℓ, and the labelling can be chosen so that $\boldsymbol{x}_1, \dots, \boldsymbol{x}_{n-\ell}$ lie in $x > 0$ and $\boldsymbol{x}_{n-\ell+1}, \dots, \boldsymbol{x}_n$ lie in $x < 0$. This information is enough to specify the symbolic description s of the orbit (note that it is NOT the rotation-compatible sequences of Section 2.4.2 as the position of the switching surface which determines s is not the same as the coding of the rotations). We shall refer to these orbits as (n, m, ℓ)-orbits.

If this periodic orbit undergoes a border collision bifurcation itself, then one point intersects the switching surface and by continuity this must be either $\boldsymbol{x}_1$ or $\boldsymbol{x}_n$ or $\boldsymbol{x}_{n-\ell}$ or $\boldsymbol{x}_{n-\ell+1}$.

Suppose that it is $\boldsymbol{x}_1$ as shown in fig. 2.15(a). Then if there is a nonsmooth saddle-node bifurcation, as suggested by the results of Section 2.7.1, the bifurcation will involve two periodic orbits: one with code s and the other with code ${}^0s$, defined to be s with the initial symbol 1 replaced by 0 whilst the rotation type of both orbits are the same. In other words, the 'partner' orbit has is a $(n, m, \ell + 1)$-orbit. Similarly, if the border collision point is $\boldsymbol{x}_{n-\ell+1}$, then it crosses at the border collision creating another code with one of the 1s in s replaced by a zero — the two orbits involved are again the (n, m, ℓ) orbit and the $(n, m, \ell+1)$-orbit as shown in fig. 2.15(d).

It turns out (see fig. 2.17) that the boundaries of the tongues are indeed generalized saddle-node bifurcations, so there is a lobe in which a (n, m, ℓ)-orbit coexists with a $(n, m, \ell + 1)$-orbit as shown in fig. 2.16 for $(n, m, \ell) = (5, 1, 3)$; note that $\ell = 2$ in fig. 2.15.

The regions (lobes) are thus defined by orbits whose ℓ description differs by one. At the shrinking point (for the piecewise affine BCNF) there is a degenerate invariant circle. This beautiful structure does not persist for typical nonlinear perturbations of the BCNF: the codimension two pinching point has a natural unfolding, see [89] for details. This is shown schematically in fig. 2.17 which combines fig. 2.16 with the change occurring through the codimension two point at which two points on the orbit are on the switching surface at the generalized saddle-node bifurcation.

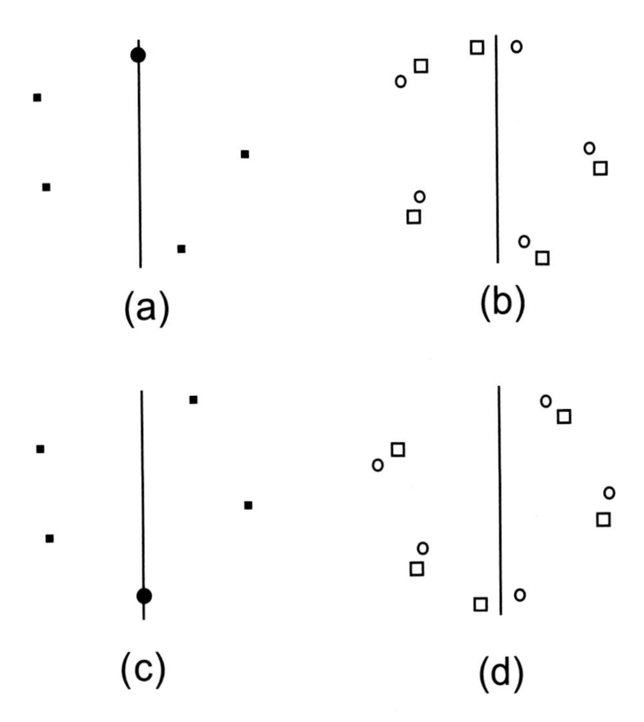

Figure 2.15: Schematic view of the generalized saddle-node bifurcations with $\ell = 2$, $m = 1$ and $n = 5$. (a) At the bifurcation parameter with $\boldsymbol{x}_1$ on the switching surface; (b) the two periodic orbits created; (c) at the bifurcation parameter with $\boldsymbol{x}_3$ on the switching surface; (d) the two orbits created by nonsmooth saddle-node bifurcation in this case.

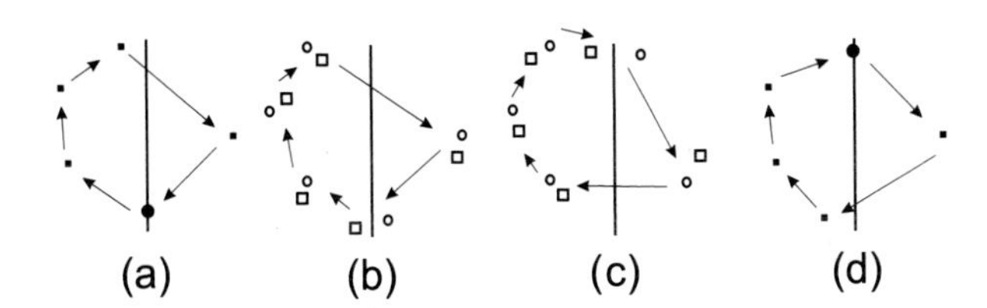

Figure 2.16: Schematic view of the generalized saddle-node bifurcations with $\ell = 3$, $m = 1$ and $n = 5$ as the parameter changes through the tongue.

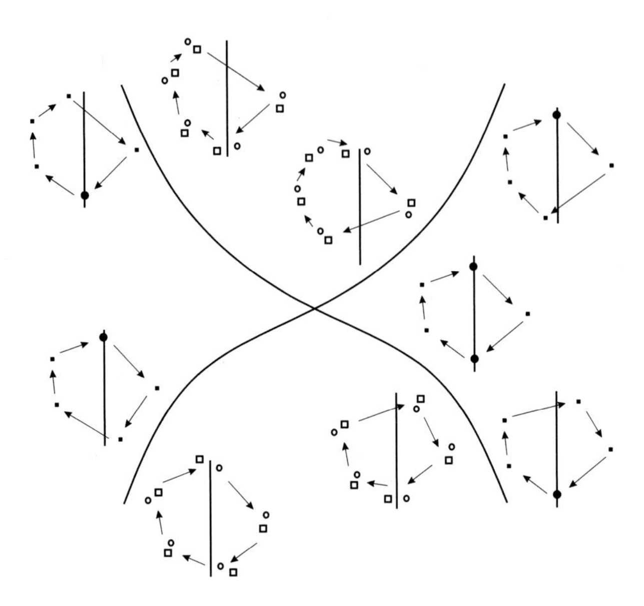

Figure 2.17: Schematic view of parameter space for the unfolding of a codimension two point at which two points on the bifurcating periodic orbit intersect the switching surface, cf. [88].

2.7.4 Infinitely many sinks

The previous section might give the impression that periodic orbits exist in splendid isolation. However, it has been recognised for many years that complicated regions of multistability exist in the border collision normal form [37]. More recently, Simpson [85] has shown that there are parameter values for the BCNF at which there are infinitely many stable periodic orbits. We will not go into the details here. Note that a similar example is explored in [24]. A piecewise-smooth (but not piecewise linear) example was described by Gambaudo–Tresser [35], and the fixed point involves in their construction satisfied the same area-preserving condition on its linearization as imposed by Simpson on the period three orbit which underpins his example. It would be interesting to understand the reason for this constraint (or convenience) in greater detail.

2.8 Robust chaos

The intersection of stable and unstable manifolds of a fixed point (a homoclinic tangle) is one of the classic mechanisms to create chaotic solutions in smooth systems. The mechanism also applies to piecewise-smooth systems, and Banerjee et al (1998) use this idea to show that there are robust chaotic attractors in the BCNF, a phenomenon they dubbed 'robust chaos'. Banerjee et al [9, 10] provide a brief plausibility argument for the proof of the chaotic attractor, here we will use results of Misieurewicz [75] which provide a more direct demonstration of the phenomenon. See [48] for a more complete discussion.

2.8.1 The Lozi map and trapping regions

Consider the restricted problem of the border collision form with

$$T_0 = -T_1 = a > 0, \quad D_0 = D_1 = -b, \quad 0 < b < 1. \tag{2.85}$$

Taking $\mu = 1$ (i.e., $\mu > 0$ by scaling) we recover the Lozi map (2.63). Note that the map is a homeomorphism and the left half plane maps to the lower half plane whilst the right half plane maps to the upper half plane. The y-axis, $x = 0$, maps to the x-axis, $y = 0$.

If the constraints of (2.85) hold and $a + b > 1$, then the system has two fixed points,

$$Y = \left(-\frac{1}{b+a-1}, -\frac{b}{b+a-1}\right), \quad X = \left(\frac{1}{1-b+a}, \frac{b}{1-b+a}\right)$$

as shown in fig. 2.18. Both are saddles; the Jacobian at Y has an stable negative eigenvalue with an eigenvector of negative slope, and an unstable positive eigenvalue with an eigenvector of positive slope whilst the Jacobian at X has an unstable negative eigenvalue with an eigenvector of negative slope, and a stable positive eigenvalue with an eigenvector of positive slope. The stable and unstable manifolds of X will be particularly important.

A great deal of the argument used to show the existence of a strange attractor for the Lozi map (2.63) relies on brute force calculation. We shall keep this to a minimum and try to emphasize the conceptual framework being developed.

The eigenvalues of the Jacobian at X are $s_\pm = \frac{1}{2}(-a \pm \sqrt{a^2 + 4b})$ with eigenvalues $\begin{pmatrix} s_\pm \\ b \end{pmatrix}$, so by a little elementary geometry the stable direction (with eigenvalue s_+) intersects the y-axis at T, where

$$T = \left(0, \frac{2b - a - \sqrt{a^2 + 4b}}{2(1 + a - b)}\right). \tag{2.86}$$

Since T is on the y-axis $f(T)$ is on the x-axis and since T is on the stable manifold of X, $f(T)$ will be the intersection of TX with the x-axis.

Similarly, the unstable direction of X intersects the (positive) x-axis at Z where

$$Z = \left(\frac{2 + a + \sqrt{a^2 + 4b}}{2(1 + a - b)}, 0\right). \tag{2.87}$$

The local unstable manifold of X thus contains the line segment $f(Z)Z$.

Since $f(Z)$ is in $x < 0$, $f^2(Z)$ lies in the lower half plane. There are thus two cases depending on whether $f^2(Z)$ lies on the left or right of the y-axis. In

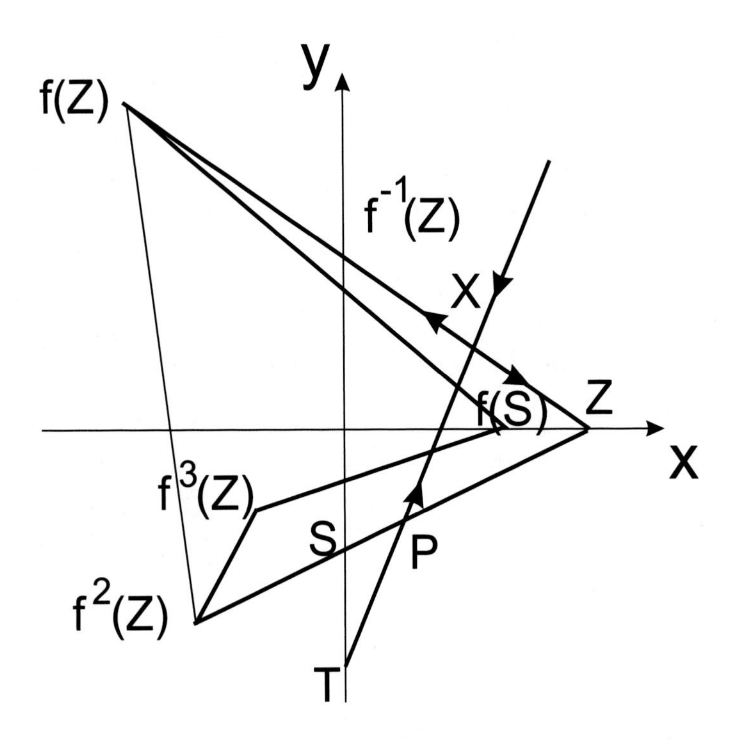

Figure 2.18: Schematic view of the bounding region and geometry of iterates for the Lozi map, after [75].

what follows below we consider only the case for which $f^2(Z)$ is on the left of the y-axis; the argument in the other case is a little more complicated (see [75]), and we leave it to the reader to find the details if they are interested.

So, by assumption (restricting the cases being considered) $f^2(Z)$ lies in the lower half plane with $x < 0$, and so $f^3(Z)$ can be calculated explicitly. This calculation can be used to show the following lemma from [75].

Lemma 2.8.1. *Consider the Lozi map* (2.63) *with parameters as described above. If $f^3(Z)$ lies in the triangle $\Delta = Zf(Z)f^2(Z)$, then $f(\Delta) \subset \Delta$.*

Proof. The geometry is shown in fig. 2.18. Let S denote the intersection of $f^2(Z)Z$ with the y-axis and note that $f(S)$ is on the x-axis to the left of Z as S lies below the origin which is below $f^{-1}(Z)$. It is an elementary calculation to show that the x-coordinate of $f^2(Z)$ is larger than that of $f(Z)$ and so the slope of $f(Z)f^2(Z)$ is negative as shown in fig. 2.18. Let $f^2(Z) = (p_1, p_2)$ and $S = (0, s_2)$. Then $p_1 < 0$ by assumption and $p_2 < s_2$ by construction. The x-coordinate of $f(S)$ is $1 + s_2$ and the x-coordinate of $f^3(Z)$ is $1 + p_2 + ap_1$ which is clearly less than $1 + s_2$ and hence $f(S)$ is to the right of $f^3(Z)$ as shown.

Δ is composed of two parts:

$$\Delta_1 = f^{-1}(Z)ZS \text{ in } x \geq 0 \quad \text{and} \quad \Delta_2 = f^{-1}(Z)f(Z)f^2(Z)S \text{ in } x \leq 0$$

(note that Δ_2 is not a triangle!). Thus $f(\Delta_1) = Zf(Z)f(S) \subset \Delta$ and $f(\Delta_2) = Zf^2(Z)f^3(Z)f(S) \subset \Delta$ and so, $f(\Delta) \subset \Delta$ as required. □

Thus Δ is a compact invariant set and hence contains an attractor provided $f^3(Z)$ is contained in Δ. Brute force calculation establishes that this is true provided a further condition is put on a and b.

Lemma 2.8.2. *If $a > 0$, $0 < b < 1$, $a > b + 1$ and $2a + b < 4$ then $f(\Delta) \subset \Delta$.*

2.8.2 Strange attractors

Banerjee et al [9, 10] provide a plausibility argument for the existence of strange attractors (albeit at different parameters of the border collision normal form, though they also discuss the case here) based on (a) the existence of transverse homoclinic intersections; and (b) the existence of heteroclinic connections between the unstable manifold of Y and the stable manifold of X. Misieurewicz [75] takes a more direct route, and whilst this is more transparent we should say something about the ideas of Banerjee et al [9, 10] before continuing.

Since X is a saddle, it has stable and unstable manifolds. Suppose that C is curve segment that crosses a part of the stable manifold of X, $W^s(X)$, transversely, then under iteration the intersection point will converge on X and the part of the remainder of the curve near the intersection point will move close to X and then expand close to the unstable manifold of X, $W^u(X)$. The Lambda Lemma [3] states that this idea can be stated precisely: in any neighbourhood of any point in $W^u(X)$ there exist a point in the image of C.

In particular, if C is itself a part of $W^u(X)$, so the intersection is a point in $W^u(X) \cap W^s(X)$, i.e., a transverse homoclinic point, then images of $W^u(X)$ lie arbitrarily close to any point in $W^u(X)$, giving a form of recurrence. Similarly, if there is a transverse intersection between $W^u(Y)$ and $W^s(X)$, then images of $W^u(Y)$ also lie arbitrarily close to any point in $W^u(X)$. Banerjee et al [9, 10] use this, together with the fact that in $x < 0$ iterates are attracted to $W^u(Y)$ and in $x > 0$ they are attracted to $W^u(X)$ to deduce that the closure of $W^u(X)$ is a chaotic invariant set.

In the case considered here we can have a transverse homoclinic point.

Lemma 2.8.3. *If S lies above T on the y-axis, then the Lozi map has a transverse homoclinic point.*

Proof. If S lies above T, then there exists an intersection point P between XT (part of the stable manifold of X) and $f^2(Z)Z$ (part of the unstable manifold of X). □

The precise condition is messy and will not be pursued here. Misieurewicz [75] proves the following.

Theorem 2.8.4. *Suppose that $a > 0$, $2a + b < 4$, $a\sqrt{2} - 2 - b > 0$ and $b < \frac{a^2-1}{2a+1}$. Then the attractor of the Lozi map (2.63) is the closure of $W^u(X)$ and the map is topologically transitive on this set.*

Remark 2.8.5. A subset $\mathcal{A}$ of $\mathbb{R}^2$ is topologically transitive if for all open U_k, $k = 0, 1$, with $U_k \cap \mathcal{A} \neq \varnothing$ there exists n such that $f^n(U_0) \cap U_1 \neq \varnothing$.

sketch. The proof is split into a number of stages which will simply be sketched here:

1. By Lemma 2.8.2, $\triangle$ contains an attracting invariant set. It is not conceptually hard (but not an easy calculation) to construct a closed set G such that $\triangle$ is contained in the interior of G and such that the attracting set

$$\tilde{G} = \bigcap_0^\infty f^n(G) = \bigcap_0^\infty f^n(\triangle) = \tilde{\triangle}.$$

   So for any $x \in \triangle$ (and in particular, for any x in the attractor) there is an open neighbourhood of x in G.

2. Let $H_0 = XZP$ and $H = \bigcup_0^\infty f^n(H_0)$. Then the boundary of H, ∂H is contained in $XP \cup W^u(X)$, $f(H) \subset H$, and $\tilde{H} = \cap f^n(H) = \tilde{\triangle}$.

3. That $\tilde{\triangle}$ is the closure of the unstable manifold of X is shown by using G and $\tilde{G}$ to show that $cl(W^u(X)) \subseteq \tilde{\triangle}$ and H and $\tilde{H}$ to show that $\tilde{\triangle} \subseteq cl(W^u(X))\tilde{\triangle}$.

4. Finally a hyperbolicity argument for expansion on the unstable manifold is used to show that f is topologically mixing on $\tilde{\triangle}$. □

2.8.3 Young's theorem

Young's theorem [101] provides an alternative approach to the chaotic attractors of border collision normal forms and their generalizations using invariant measures. This is not the place to give a detailed technical description of the theorem, but it is nonetheless useful to know that such techniques exist and can be applied to examples.

A measure μ on a space is essentially a way of assigning size or probability to subsets (strictly speaking, measureable subsets) of the space. Thus if X is a compact subset of the plane, a (probability) measure is a map from (measureable) subsets U of X to the $[0,1]$ such that

(i) $\mu(\varnothing) = 0$, $\mu(X) = 1$,

(ii) $\mu(U \cup V) \leq \mu(U) + \mu(V)$ with equality if $U \cap V = \varnothing$,

and a measure is an invariant measure of a map $f: X \to X$ if for all $U \subseteq X$

$$\mu(f^{-1}(U)) = \mu(U).$$

Invariant measures provide ways of linking spatial and temporal averages: if $g: X \to \mathbb{R}$ is a nice (integrable) function then we would like a result of the form

$$\frac{1}{n}\sum_{0}^{n-1} g(f^n(x)) \to \int_X g d\mu$$

as $n \to \infty$ (for μ almost all x). This is true for *ergodic* measures: i.e., invariant probability measures with the property that for every invariant set E (i.e., measureable sets with $f^{-1}(E) = E$) either $\mu(E) = 0$ or $\mu(E) = 1$.

Young's theorem provides a way of proving that nice measures exist for robust chaos.

Let $R = [0,1] \times [0,1]$ and let $S = \{a_1, \ldots, a_k\} \times [0,1]$ be a set of vertical switching surfaces with $0 < a_1 < \cdots < a_k < 1$. Then $f: R \to R$ is a Young map if f is continuous, f and its inverse are C^2 on $R \backslash S$ and $f = (f_1, f_2)^T$ satisfies the expansion properties (H1)–(H3) below on $R \backslash S$.

$$(H1) \quad \inf\left\{\left(\left|\frac{\partial f_1}{\partial x}\right| - \left|\frac{\partial f_1}{\partial y}\right|\right) - \left(\left|\frac{\partial f_2}{\partial x}\right| - \left|\frac{\partial f_2}{\partial y}\right|\right)\right\} \geq 0,$$

$$(H2) \quad \inf\left(\left|\frac{\partial f_1}{\partial x}\right| - \left|\frac{\partial f_1}{\partial y}\right|\right) = u > 1, \quad \text{and}$$

$$(H3) \quad \sup\left\{\left(\left|\frac{\partial f_1}{\partial y}\right| + \left|\frac{\partial f_2}{\partial y}\right|\right)\left(\left|\frac{\partial f_1}{\partial x}\right| - \left|\frac{\partial f_1}{\partial y}\right|\right)^{-2}\right\} < 1.$$

Young's theorem describes measures that project nicely onto one dimension. Technically this is expressed as having absolutely continuous conditional measures on unstable manifolds. Intuitively this means that the measure projects nicely onto one dimension.

Let $\mathrm{Jac}(f)$ denote the Jacobian matrix of f and recall that u is defined in (H2).

Theorem 2.8.6 (Young, [101]). *If f is a Young map, $|\mathrm{Jac}(f)| < 1$ for $x \in R \backslash S$, and there exists $N \geq 1$ s.t. $u^N > 2$ and if $N > 1$, then $f^k(S) \cap S = \varnothing$, $1 \leq k < N$, then f has an invariant probability measure that has absolutely continuous conditional measures on unstable manifolds.*

Since the result is for piecewise C^2 maps and the conditions only depend on derivatives this result has the important corollary that results for the piecewise linear border collision normal form, which should more correctly be called a truncated normal form, persist when small nonlinear terms are added.

Historical note. The theorem as actually published [101] has $u^N > 2$ and $f^k(S) \cap S = \varnothing$, $1 \leq k \leq N$ (note the non-strict inequality in the last expression). However, no extra conditions on images of S are required if $N = 1$ and if $N > 1$, then the requirement is that f^N has similar geometry on vertical strips, which only requires non-intersection up to the $(N-1)^{th}$ iterate, so we are confident that Theorem 2.8.6 is what was intended.

The criteria for the theorem to hold are easy to verify numerically making it possible to determine regions on which Young's theorem holds and compare these with theoretical bounds in [10]; see [43] for details.

2.9 Two-dimensional attractors

The BCNF can also have robust two-dimensional attractors. These results use some beautiful theory for general piecewise linear maps due to Buzzi and Tsujii. These will be described in the second section — first we describe another context in which the existence of two-dimensional attractors can be deduced from first principles. Note that the existence of two-dimensional attractors implies expansion in all directions, so the only way this can occur is through folding, i.e., the map must be non-invertible: $D_0 D_1 < 0$.

2.9.1 A Markov partition

In Section 2.2.1 we saw that Markov partitions and their associated graphs provide a good way to analyze dynamics. The idea in this section is to construct an example with a two-dimensional Markov partition and then show that the map (or an iterate of the map) is uniformly expanding on each region defining the Markov partition.

Consider the BCNF with $D_0 < 0$, $D_1 > 0$ and $\mu = 1$. We shall start by constructing a simple bounding region and then try to describe the dynamics in this region. Note that the conditions on D_k, $k = 0, 1$, imply that the images of both the left and the right half planes map to the lower half plane.

Let $O = (0,0)$ so $P_1 = f(0,0) = (1,0)$. Suppose that $P_2 = f(P_1)$ is in $x > 0$ and $P_3 = f(P_2)$ lies on the y-axis. so $P_4 = f(P_3)$ lies on the x-axis and we shall assume this can be chosen so that P_4 is in the left half plane as shown in fig. 2.19(a).

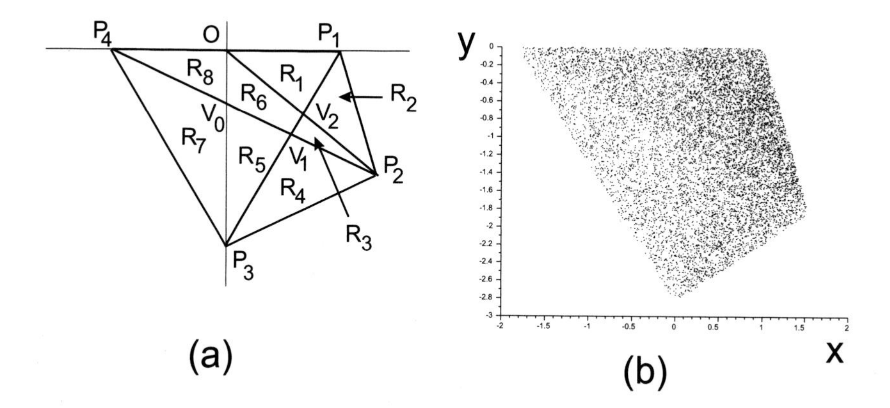

Figure 2.19: (a) Schematic view of the Markov partition; (b) numerical solution for parameters given below.

To achieve this will require only one real constraint (that P_3 lies the y-axis), the remainder are open conditions.

Next, choose the parameters such that $f(P_4) = P_2$ (two conditions; these will fix T_0 and D_0) and finally arrange it so that the straight line P_4P_2 intersects the y-axis at $V_0 = (0, -1)$, the preimage of O (one real condition). This gives four real conditions for the four parameters $T_k,\ D_k,\ k = 0, 1$. We will show that these can be solved below, but before verifying this let us consider the consequences (see fig. 2.19a again).

Let V_1 be the intersection of P_1P_3 with V_0P_2, so its image will lie on the intersection of P_2P_4 and OP_3, i.e., $f(V_1) = V_0$. Similarly let V_2 be the intersection of P_1P_3 and OP_2 so $f(V_2) = V_1$. The lines connecting the points O, $P_1, \ldots, P_4$, V_0, V_1 and V_2 divide the trapping region $OP_1P_2P_3P_4$ into eight sectors

$$\begin{array}{llll} R_1 = OV_2P_1, & R_2 = P_1V_2P_2, & R_3 = P_2V_1V_2, & R_4 = P_2V_1P_3, \\ R_5 = P_3V_1V_0, & R_6 = OV_0V_1V_2, & R_7 = P_3V_0P_4, & R_8 = P_4V_0O. \end{array} \tag{2.88}$$

These have been chosen so that

$$\begin{array}{lll} f(R_1) = R_2 \cup R_3, & f(R_2) = R_4, & f(R_3) = R_5, \\ f(R_4) = R_4 \cup R_7, & f(R_5) = R_8, & f(R_6) = R_1 \cup R_6, \\ f(R_7) = R_3 \cup R_6 \cup R_8, & f(R_8) = R_1 \cup R_2. & \end{array} \tag{2.89}$$

This is therefore a two-dimensional Markov partition and the symbolic description of orbits is easy to describe using a Markov graph in precisely the same way as in Section 2.2.1. A little more work is required to show that the map is transitive on the invariant region, see [52] for details.

Let us check that this is possible. By direct calculation,

$$P_2 = (T_1 + 1, -D_1), \quad P_3 = (T_1(T_1 + 1) - D_1 + 1, -D_1(T_1 + 1)),$$

and hence the first constraint is that

$$D_1 = T_1(T_1 + 1) + 1. \tag{2.90}$$

In this case set $t = T_1$ so $D_1 = t^2 + t + 1$ and

$$P_3 = (0, -D_1(t^2 + t + 1)), \quad P_4 = (1 - D_1(t + 1), 0),$$

and P_4 is in $x < 0$ provided $D_1(t + 1) > 1$ and note that this is certainly true if $D_1 > 1$ and $t > 0$. Now the line P_2P_4 intersects the y-axis at $V_2 = (0, -1)$ if (by similar triangles)

$$\frac{1}{D_1(t+1) - 1} = \frac{D_1}{D(t+1) + t}$$

and after a little algebra (involving factorization of a quintic in t) this holds if

$$t^3 + t^2 + t - 1 = 0, \quad D_1 = \tfrac{1}{t}. \tag{2.91}$$

A simple root-finding method shows that this has a positive solution with $T_1 = t \approx 0.543689$, $D_1 \approx 1.839287$ and solving the equations for T_0 and D_0 gives $T_0 = -t^2 \approx -0.295598$, $D_0 = -1$.

Figure 2.19(b) shows a numerically calculated solution for these parameter values. Glendinning–Wong [52] show that an expansion condition holds on iterates of the map which implies transitivity on the whole region $OP_1P_2P_3P_4$. They also derive conditions for a sequence of other parameters having a similar Markov property.

2.9.2 Piecewise linear maps

A number of general results were proved in around 2000 proving the existence of two-dimensional attractors for piecewise linear maps. These all rely on expansion of each individual map, but the technical assumptions are more general than the BCNF as continuity across boundaries is not assumed. Here we follow Buzzi [14] and Tsujii [94].

Let $\mathcal{D}$ be a polygonal region in $\mathbb{R}^2$, i.e., a compact connected region whose boundary is a finite union of straight line segments. Let $\mathcal{P}$ be a finite collection of non-intersecting open polygonal regions $\{P_i\}_{i=1}^m$ such that the union of the closures of these polygons is $\mathcal{D}$. Then a map $F: \bigcup P_i \to \mathcal{D}$ is a *piecewise affine map* if $F|_{P_i}$ is an affine map, $i = \{1, \dots, m\}$. If in addition there exists $\lambda > 1$ and a metric $d: \mathbb{R}^2 \to \mathbb{R}$ such that for each $i \in \{1, \dots, m\}$ $F|_{P_i}$ is expanding, i.e.,

$$d(F(x), F(y)) \geq \lambda d(x, y) \quad \text{for all } x \in P_i$$

$i = 1, \dots, m$, then F is a *piecewise expanding affine map*. The main result that can be applied to the BCNF shows that there are two-dimensional attractors. Like Young's theorem it uses the idea of invariant measures to describe the dynamics, but it is the existence of open sets in the attractor which implies that the attractor has topological dimension two rather than simply Hausdorff dimension equal to two.

Theorem 2.9.1 (Buzzi, [13, 14]; Tsujii, [94]). *Suppose F is a piecewise expanding affine map of a planar polygonal region $\mathcal{D}$. Then there exists an attractor in $\mathcal{D}$ such that F has an absolutely continuous invariant measure on the attractor and the attractor contains open sets.*

Unfortunately, the BCNF is not expanding (at least in the standard Euclidean metric), so a little more work needs to be done in order to apply this result.

2.9.3 Robust bifurcations to two-dimensional attractors

The examples of Section 2.9.1 can be proved to have two-dimensional attractors, but they exist at special values of the parameters. The results of Buzzi and Tsujii of Section 2.9.2 make it possible to prove the existence of such sets for open sets of parameters. It is even possible to construct open conditions so that the border collision bifurcation has a stable fixed point if $\mu < 0$ and a two-dimensional attractor if $\mu > 0$ [47]. The proof follows the rather easier path of [46]. An example of a two-dimensional attractor with

$$T_L = -0.1, \quad D_L = -8/11, \quad T_R = 0.05, \quad D_R = 1.99,$$

and $\mu = 1$ is given in fig. 2.20.

Theorem 2.9.2 (Glendinning, [47]). *There exists an open region $\mathcal{D} \subset \mathbb{R}^4$ such that if $(T_0, D_0, T_1, D_1) \in \mathcal{D}$, then the BCNF* (2.64), (2.65) *has a stable fixed point if $\mu < 0$ and a fully two-dimensional attractor if $\mu > 0$.*

Proof. From the results of Section 2.7.1 the choices $1+D_0-T_0 > 0$ and $1+D_1-T_1 > 0$ imply that there is a fixed point in $x < 0$ if $\mu < 0$ and a fixed point in $x > 0$ if $\mu > 0$. The fixed point in $x < 0$ is stable (when it exists) provided the eigenvalues of the Jacobian have modulus less than one, i.e., if

$$|D_0| < 1 \quad \text{and} \quad |T_0| < 1 + D_0. \tag{2.92}$$

Now consider $\mu > 0$, so by scaling we can assume that $\mu = 1$. The pattern will be similar to proofs of Section 2.8: we begin by constructing an absorbing region for well-chosen parameters. Fix $\epsilon > 0$ (to be chosen small enough later) and suppose that

$$|T_k| < \epsilon, \; k = 0, 1, \quad -D_0 \in (\frac{6}{11}, \frac{10}{11}), \quad D_1 \in (2-\epsilon, 2). \tag{2.93}$$

Clearly (2.92) is satisfied for small ϵ, so if $\mu < 0$, there is a stable fixed point. If $\mu = 1$, consider the rectangular region with

$$-(1 + 0.05 - 4\epsilon) \le x \le 1 + 4\epsilon, \quad -(2 + 0.05 - 2\epsilon) \le y \le 2\epsilon. \tag{2.94}$$

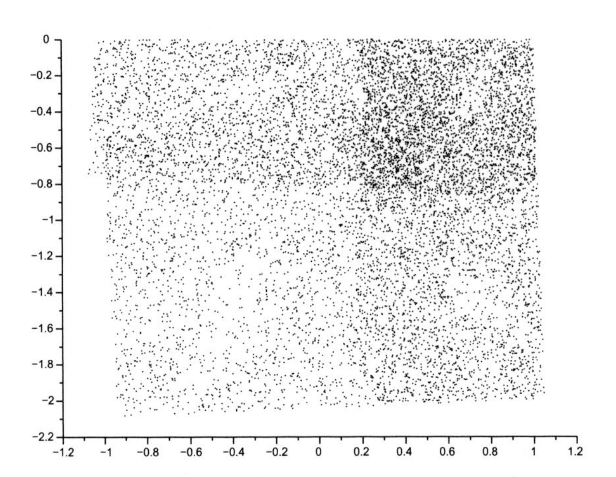

Figure 2.20: Numerically calculated attractor for the BCNF with parameters as given in the text.

If (x, y) is in this rectangle, then the image is (x', y') with $x' = 1 + y + T_k x$ and so taking maximum and minimum values

$$1 - (2 + 0.05 - 2\epsilon) - \epsilon(1 + 0.05 - 4\epsilon) \leq x' \leq 1 + 2\epsilon + \epsilon(1 + 4\epsilon),$$

i.e., $-(1 + 0.05 - c_1\epsilon - 4\epsilon^2) \leq x' \leq 1 + 3\epsilon + 4\epsilon^2$ and so, provided ϵ is sufficiently small, x' satisfies the same rectangle constraint as x in (2.94).

Similarly, $y' = -D_0 x$ if $x < 0$, so y' is is negative in this case and takes a minimum value of around $-\frac{10}{11}$ which is small in modulus compared with the boundary of the rectangle and so y' comfortably satisfies the constraints of the rectangle for ϵ small. If $x > 0$ then $y' = -D_1 x$ and so again y' is negative and

$$-2(1 + 4\epsilon) \leq y'.$$

Hence, provided $0.05 - 2\epsilon > 8\epsilon$ this will again lie in the region defined by (2.94). Thus for small enough $\epsilon > 0$ the region (2.94) is invariant.

To prove expansion and hence apply results of Section 2.9.2, we need to know a little more about the dynamics in this region.

Suppose (x, y) lies in the rectangle defined by (2.94) with $x < 0$. Then the image point (x', y') has $y' = -D_0 x < 0$ and hence the second iterate will have x-coordinate less than $1 - D_0 x + \epsilon|x'|$ which is greater than zero for sufficiently small ϵ as the maximum of x is close to 1.05 so $|D_0 x| \leq \frac{21}{22}$ up to terms of order ϵ. Thus if $x < 0$, the x-coordinate of $f^2(x, y)$ is in $x > 0$.

Note that the linear matrices of the BCNF with $|T_k| \approx 0$ have the form $\begin{pmatrix} 0 & 1 \\ \alpha & 0 \end{pmatrix}$, $\alpha \in \{-D_0, -D_1\}$, suppose we multiply four of these together with $\alpha_1, \ldots, \alpha_4$ as the bottom left coefficients. Straightforward calculation show that we obtain

$$\begin{pmatrix} \alpha_2\alpha_4 & 0 \\ 0 & \alpha_1\alpha_3 \end{pmatrix}.$$

Now, the Jacobian of f^4, Df^4, is just a product of BCNF matrices along the orbit, so if $\alpha_1 = -D_0$, then the second iterate is in $x > 0$ and so $\alpha_3 = -D_1$ and similarly for α_2. Thus the only combinations possible are D_1^2 which is close to 4, and $D_0 D_1$ which is close to $-\frac{12}{11}$ or larger. Adding in the order ϵ corrections will not change the fact that the Jacobian of f^4 is expanding and hence f^4, defined on regions on which it is linear, is an expanding piecewise linear map and has a two-dimensional attractor by Theorem 2.9.1. It is straightforward to show that this implies that f itself has a two-dimensional attractor and the result is proved. □

2.10 Challenges

There are many possible generalizations of th results presented here, and other directions that could have been taken. Here we mention just a few.

2.10.1 Other classes of maps

In Sections 2.1.3 and 2.2.3 we described Nordmark's square root map [78]. Square root maps appear in many contexts in piecewise-smooth systems [21] and so it would be natural to put more attention into the phenomena that can arise in these cases (e.g., [7]). Once again though, the issue should be to understand what can be said usefully. It may be that the classes are too large, or the bifurcation phenomena too complicated, to give complete descriptions and therefore the skill is to find useful but finite statements: less is more (cf. Section 2.1.4).

The square root map introduces a particular singularity in the derivative of the map. But in the piecewise-smooth world it is always possible (at least in principle) to introduce more discontinuities. When is this useful? When is it interesting? What about infinitely many discontinuities? Mathematicians can always think of generalizations, but it is probably best (in general) to allow applications to suggest what is most worthwhile.

The work on the border collision normal form uses the fact that the map is piecewise linear in a number of ways: it means quite a lot of features can be computed by brute force (Section 2.7.2 for example) and it means that iterates of straight lines are straight lines, simplifying geometric arguments considerably (this is key to Buzzi's proofs for piecewise expanding maps in Section 2.9.2). However, apart from Young's theorem (Section 2.8.3) and the original robust chaos argument of [10] relatively few results seem to carry over easily. The effect of nonlinear terms, and more generally higher-order terms in normal forms, seems an important topic for future research.

The final area, and the one which will occupy the remainder of these lectures, is the effect of higher dimensions. As argued in [49, 50, 51] the number of cases can multiply hugely as the dimension of the phase space increases, but there are still examples of results that are either independent of the dimension.

2.10.2 Higher dimensions: periodic orbits

In Section 2.7.1 it was possible to compute precise criteria for the existence of fixed point and orbits of period two for the border collision normal form in two dimensions, and to give criteria for their stability. This is also possible for the BCNF in $\mathbb{R}^n$, where the normal form is (2.64) with constant $\mu(1,0,\dots,0)^T$ and the matrices A_0 and A_1 are in observer canonical form [19]

$$A_k = \begin{pmatrix} r_{k1} & 1 & 0 & 0 & \cdots \\ r_{k2} & 0 & 1 & 0 & \cdots \\ r_{k3} & 0 & 0 & 1 & \cdots \\ \vdots & . & . & . & \cdots \\ r_{kn} & 0 & 0 & \cdots & 0 \end{pmatrix}, \quad k = 0, 1. \tag{2.95}$$

Without going through the details, we will state the result, which depends on the *index* of the matrices A_0 and A_1.

Definition 2.10.1. The index σ_k^+ (resp., σ_k^-) of the matrix A_k of (2.95) is defined by the number of real eigenvalues of A_k greater than 1 (resp., less than 1), $k = 0, 1$.

The index gives information about the fixed points and points of period two [22, 87].

Theorem 2.10.2. *Consider the BCNF in $\mathbb{R}^n$. Let $\boldsymbol{x}_k$ denote a fixed point of the BCNF in $x < 0$ if $k = 0$ and $x > 0$ if $k = 1$. Then,*

(i) *if $\sigma_0^- + \sigma_1^-$ is even and $\sigma_0^+ + \sigma_1^+$ is even, then $\boldsymbol{x}_0$ and $\boldsymbol{x}_1$ exist for different signs of μ and there are no period two orbits if $\mu \neq 0$;*

(ii) *if $\sigma_0^- + \sigma_1^-$ is even and $\sigma_0^+ + \sigma_1^+$ is odd, then $\boldsymbol{x}_0$ and $\boldsymbol{x}_1$ exist for the same sign of μ and there are no period two orbits if $\mu \neq 0$;*

(iii) *if $\sigma_0^- + \sigma_1^-$ is odd and $\sigma_0^+ + \sigma_1^+$ is even, then $\boldsymbol{x}_0$ and $\boldsymbol{x}_1$ exist for different signs of μ and an orbit of period two orbits exists for one sign of μ;*

(iv) *if $\sigma_0^- + \sigma_1^-$ is odd and $\sigma_0^+ + \sigma_1^+$ is odd, then $\boldsymbol{x}_0$ and $\boldsymbol{x}_1$ and an orbit of period two orbits exists for one sign of μ.*

If $\boldsymbol{x}_0$ and $\boldsymbol{x}_1$ both exist for the same sign of μ, then $\sigma_0^+ + \sigma_1^+$ is odd and so at least one of them is non-zero. Hence at least one of the matrices A_0 and A_1 has an eigenvalue with modulus greater than 1.

Corollary 2.10.3. *The BCNF cannot have coexisting stable fixed points.*

In fact, with a little more work it can be shown that if the period two orbit if it is stable then the fixed point that coexists with it is unstable [87].

Most of the analysis of Section 2.7.2 was actually independent of the dimension of phase space, so the analysis can be used to describe periodic orbits, mode locking and shrinking points in higher dimensional systems; see [87] for details.

2.10.3 Higher dimensions: n-dimensional attractors

The results of Buzzi and Tsujii described in Section 2.9.2 hold in $\mathbb{R}^n$, $n > 2$, but with a slight caveat: the attractors may not have topological dimension n, i.e., they may not contain open sets, though they always have Hausdorff dimension n and topological dimension n on a generic set of parameters. This makes it possible to prove results analogous to Theorem 2.9.2 but with that technical restriction.

Theorem 2.10.4 (Glendinning, [46]). *There exists an open set $U \subset \mathbb{R}^{2n}$ such that if $(r_{01}, \ldots, r_{0n}, r_{11}, \ldots, r_{1n}) \in U$, then the border collision normal form in $\mathbb{R}^n$ with matrices (2.95) has a stable fixed point if $\mu < 0$ and an attractor with Hausdorff dimension equal to n if $\mu > 0$. This attractor has topological dimension equal to n generically in U.*

It appears harder to generalize Young's results of Section 2.8.3, though a recent result of Zhang [102] extends her result to $\mathbb{R}^3$ with two-dimensional unstable manifolds. It would be very interesting to see this extended to higher dimension, and higher dimensional unstable manifolds.

Bibliography

[1] M.A. Aizerman and F.R. Gantmakher, On the stability of equilibrium positions in discontinuous systems, *Prikl. Mat. i Mekh.* **24** (1960), 283–93.

[2] M.A. Aizerman and E.S. Pyatnitskii, Fundamentals of the theory of discontinuous systems I,II, *Automation and Remote Control* **35** (1974), 1066–79, 1242–92.

[3] K.T. Alligood, T.D. Sauer, and J.A. Yorke, *Chaos: An introduction to dynamical systems*, Springer, 1996.

[4] L. Alsedà, J. Llibre, M. Misieurewicz, and C. Tresser, Periods and entropy for Lorenz-like maps, *Ann. de l'Inst. Fourier* **39** (1989), 929–952.

[5] A.A. Andronov, A.A. Vitt, and S.E. Khaikin, *Theory of oscillations*, Moscow: Fizmatgiz (in Russian), 1959.

[6] V.I. Arnold, V.V. Goryunov, O.V. Lyashko and V.A. Vasiliev, *Dynamical systems VI: Singularity theory I. Local and global theory*, Springer-Verlag, 1993.

[7] V. Avrutin, P.S. Dutta, M. Schanz, and S. Banerjee, Influence of a square-root singularity on the behavior of piecewise-smooth maps, *Nonlinearity* **23** (2010), 445–463.

[8] V. Avrutin, M. Schanz, and S. Banerjee, Multi-parametric bifurcations in a piecewise-linear discontinuous map, *Nonlinearity* **19** (2006), 1875–1906.

[9] S. Banerjee and C. Grebogi, Border collision bifurcations in two-dimensional piecewise smooth maps, *Phys. Rev. E* **59** (1999), 4052–4061.

[10] S. Banerjee, J.A. Yorke, and C. Grebogi, Robust chaos, *Phys. Rev. Lett.* **80** (1998), 3049–3052.

[11] D. Berry, *Nonwandering sets of Lorenz maps*, Ph.D. Thesis, University of Warwick, 1999.

[12] B. Brogliato, *Impact in mechanical systems – analysis and modelling*, Springer-Verlag, 2000.

P. Glendinning, M. R. Jeffrey, *An Introduction to Piecewise Smooth Dynamics*, Advanced Courses in Mathematics - CRM Barcelona, https://doi.org/10.1007/978-3-030-23689-2

[13] J. Buzzi, Absolutely continuous invariant measures for generic multi-dimensional piecewise affine expanding maps, *Int. J. Bifn. & Chaos* **9** (1999), 1743–1750.

[14] J. Buzzi, Thermodynamic formalism for piecewise invertible maps: Absolutely continuous invariant measures as equilibrium states, in: *Smooth Ergodic Theory and Its Applications* (A. Katok, R. de la Llave, Y. Pesin, and H. Weiss, eds), AMS Proc. Symp. Pure Math. **69**, pp. 749–784, AMS, 2001.

[15] A. Colombo, M.R. Jeffrey, J.T.Lázaro, J.M. Olm (eds.), *Extended Abstracts Spring 2016. Nonsmooth Dynamics*, Trends in Mathematics: Research Perspectives CRM Barcelona, Birkhäuser, 2016.

[16] P. Collet and J.P. Eckmann, *On iterated maps of the interval as dynamical systems*, Birkhäuser, 1980; reprinted in *Modern Birkhäuser Classics*, 2009.

[17] W. de Melo and S. van Strien, *One-dimensional dynamics*, Springer, Berlin, 1993.

[18] R. Devaney, *An introduction to chaotic dynamical systems (second edition)*, Addison-Wesley, Redwood City, 1989.

[19] M. di Bernardo, Normal forms of border collision in high dimensional non-smooth maps, *Proceedings IEEE ISCAS 2003* **3** (2003), 76–79.

[20] M. di Bernardo, C. Budd and A.R. Champneys, Corner collision implies border-collision bifurcation, *Physica D* **154** (2001), 171–194.

[21] M. di Bernardo, C. Budd, A.R. Champneys, and P. Kowalczyk, *Piecewise-smooth dynamical systems: Theory and applications*, Applied Mathematical Sciences, Vol. 163, Springer, London, 2008.

[22] M. di Bernardo, M.I. Feigin, S.J. Hogan, and M.E. Homer, Local analysis of C-bifurcations in n-dimensional piecewise-smooth dynamical systems, *Chaos, Solitons & Fractals* **10** (1999), 1881–1908.

[23] M. di Bernardo, U. Montanaro, and S. Santini, Canonical forms of generic piecewise linear continuous systems, *IEEE Trans. Automatic Control* **56** (2011), 1911–1915.

[24] Y. Do and Y.-C. Lai, Multistability and arithmetically period-adding bifurcations in piecewise smooth dynamical systems, *Chaos* **18** (2008), 043107.

[25] V.A. Dobrynskiy, On attractors of piecewise linear 2-endomorphisms, *Nonlinear Anal.* **36** (1999), 423–455.

[26] R. Edwards, A. Machina, G. McGregor, and P. van den Driessche, A modelling framework for gene regulatory networks including transcription and translation, *Bull. Math. Biol.* **77** (2015), 953–983.

[27] C.P. Fall, E.S. Marland, J.M. Wagner, and J.J. Tyson, *Computational cell biology*, Springer-Verlag 2002.

[28] A.F. Filippov, Differential equations with discontinuous right-hand side, *Amer. Math. Soc. Transl. Ser. 2* **42** (1964), 19–231.

[29] A.F. Filippov, *Differential equations with discontinuous righthand sides*, Kluwer Academic Publ. Dortrecht, 1988 (Russian 1985).

[30] I. Flügge-Lotz, *Discontinuous automatic control*, Princeton University Press, 1953.

[31] J.M. Gambaudo, *Ordre, désordre, et frontière des systèmes Morse–Smale*, Thesis, Université de Nice, 1987.

[32] J.M. Gambaudo, P. Glendinning, and C. Tresser, The gluing bifurcation: I. symbolic dynamics of the closed curves, *Nonlinearity* **1** (1986), 203–214.

[33] J.M. Gambaudo, P. Glendinning, and C. Tresser, Stable cycles with complicated structure, in: *Instabilities and Nonequilibrium Structures* (E. Tirapegui and D. Villarroel, eds.), Reidel, Dordrecht, 1987.

[34] J.M. Gambaudo, I. Procaccia, S. Thomae, and C. Tresser, New universal scenarios for the onset of chaos in Lorenz type flows, *Phys. Rev. Lett.* **57** (1986), 925–928.

[35] J.M. Gambaudo and C. Tresser, Simple models for bifurcations creating horseshoes, *J. Stat. Phys.* **32** (1983), 455–476.

[36] J.M. Gambaudo and C. Tresser, Dynamique regulière ou chaotique. Applications du cercle ou de l'intervalle ayant une discontinuité, *C.R. Acad. Sci. (Paris) Série I* **300** (1985), 311–313.

[37] L. Gardini, Some global bifurcations of two-dimensional endomorphisms by use of critical lines, *Nonlinear Anal.* **18** (1992), 361–399.

[38] L. Gardini, V. Avrutin, and I. Sushko, Codimension-2 border collision bifurcations in one-dimensional discontinuous piecewise-smooth maps., *Int. J. Bif. & Chaos* **24** (2014), 1450024.

[39] L. Gardini, R. Makrooni, and I. Sushko, Cascades of alternating smooth bifurcations and border collision bifurcations in a family of discontinuous linear-power maps, *Discrete & Cont. Dynam. Syst. B* **23** (2018), 701–729.

[40] P. Glendinning, *Homoclinic bifurcations in ordinary differential equations*, a Fellowship Dissertation, King's College, Cambridge, 1985.

[41] P. Glendinning, The anharmonic route to chaos: kneading theory, *Nonlinearity* **6** (1993), 349–376.

[42] P. Glendinning, *Stability, instability and chaos*, Cambridge Univ. Press, 1994.

[43] P. Glendinning, Invariant measures for the border collison normal form, MIMS preprint, Manchester, 2011.

[44] P. Glendinning, Attractors with dimension n for open sets of parameter space in the n-dimensional border collision normal form, *Int. J. Bif. & Chaos* **24** (2014), 1450164.

[45] P. Glendinning, Renormalization for the boundary of chaos in piecewise monotonic maps with a single discontinuity, *Nonlinearity* **27** (2014), R143–R162.

[46] P. Glendinning, Bifurcation from stable fixed point to N-dimensional attractor in the border collision normal form, *Nonlinearity* **28** (2015), 3457–3464.

[47] P. Glendinning, Bifurcation from stable fixed point to two-dimensional attractor in the border collision normal form, *IMA J. Appl. Math.* (2016). doi: 10.1093/imamat/hxw001.

[48] P. Glendinning, Robust chaos revisited, *European Physical Journal Special Topics* **226** (2017), 1721. doi:10.1140/epjst/e2017-70058-2.

[49] P. Glendinning, Less is more I: a pessimistic view of piecewise-smooth bifurcation theory, in: *Extended Abstracts Spring 2016. Nonsmooth Dynamics* (A. Colombo, M.R. Jeffrey, J.T. Lazaro and J.M. Olm, eds.), Research Perspectives CRM Barcelona, pp. 71–75, Birkhäuser, 2017.

[50] P. Glendinning, Less is more II: an optimistic view of piecewise-smooth bifurcation theory, in: *Extended Abstracts Spring 2016. Nonsmooth Dynamics* (A. Colombo, M.R. Jeffrey, J.T. Lazaro and J.M. Olm, eds.), Research Perspectives CRM Barcelona, pp. 77–81, Birkhäuser, 2017.

[51] P. Glendinning and M.R. Jeffrey, Grazing-sliding bifurcations, border collision maps and the curse of dimensionality for piecewise-smooth bifurcation theory, *Nonlinearity* **28** (2015), 263–283.

[52] P. Glendinning and C.H. Wong, Two-dimensional attractors in the border-collision normal form, *Nonlinearity* **24** (2011), 995–1010.

[53] A. Granados, L. Alsedà, and M. Krupa, The period adding and incrementing bifurcations: from rotation theory to applications (2015). arXiv:1407.1895v3.

[54] J. Guckenheimer and R.F. Williams, Structural stability of Lorenz attractors, *Publ. Math. IHES* **50** (1979), 59–72.

[55] N. Guglielmi and E. Hairer, Classification of hidden dynamics in discontinuous dynamical systems, *SIADS* **14**(3) (2015), 1454–77.

[56] S. Ito, H. Nakada, and S. Tanaka, On unimodal linear transformations and chaos, *Proc. Japan Acad. Ser. A Math. Sci.* **55** (1979), 231–236.

[57] S. Ito, H., and S. Tanaka, On unimodal linear transformations and chaos II, *Tokyo J. Math.* **2** (1979), 241–259.

[58] M.R. Jeffrey, Non-determinism in the limit of nonsmooth dynamics, *Physical Review Letters* **106** no. 25 (2011), 254103.

[59] M.R. Jeffrey, Hidden dynamics in models of discontinuity and switching, *Physica D* **273** no. 4 (2013), 34–45.

[60] M.R. Jeffrey, Dynamics at a switching intersection: hierarchy, isonomy, and multiple-sliding, *SIADS* **13** no. 3 (2014), 1082–1105.

[61] M.R. Jeffrey, Exit from sliding in piecewise-smooth flows: deterministic vs. determinacy-breaking, *Chaos* **26** no. 3 (2016), 033108:1–20.

[62] M.R. Jeffrey, Hidden degeneracies in piecewise-smooth dynamical systems, *Int. J. Bif. Chaos* **26** no. 5 (2016), 1650087:1–18.

[63] M.R. Jeffrey, Why nonsmooth?, in: *Extended Abstracts Spring 2016. Nonsmooth Dynamics*, Trends in Mathematics: Research Perspectives CRM Barcelona, pp. 101–105, Birkhäuser, 2016.

[64] M.R. Jeffrey, An update on that singularity. in: *Extended Abstracts Spring 2016. Nonsmooth Dynamics*, Trends in Mathematics: Research Perspectives CRM Barcelona, pp. 107–112, Birkhäuser, 2016.

[65] L. Jonker and D.A. Rand, Bifurcations in one-dimension. I. The nonwandering set, *Invent. Math.* **62** (1981), 347–365.

[66] J.P. Keener, Chaotic behavior in piecewise continuous difference equations, *Trans. Amer. Math. Soc.* **261** (1980), 589–604.

[67] Yu.A. Kuznetsov, S. Rinaldi, and A. Gragnani, One-parameter bifurcations in planar Filippov systems, *Int. J. Bif. Chaos* **13** (2003), 2157–2188.

[68] V. Kulebakin, On theory of vibration controller for electric machines, *Theor. Exp. Electon* (in Russian) **4** (1932).

[69] Y.A. Kuznetsov, *Elements of applied bifurcation theory*, Springer, 2004.

[70] T.-Y. Li and J.A. Yorke, Period three implies chaos, *Am. Math. Monthly* **82** (1975), 985–992.

[71] J. Llibre, D.N. Novaes and M.A. Teixeira, Maximum number of limit cycles for certain piecewise linear dynamical systems, *Nonlinear Dynamics* **82** no. 3 (2015), 1159–1175.

[72] A. Machina, R. Edwards, and P. van den Dreissche, Singular dynamics in gene network models, *SIADS* **12** no. 1 (2013), 95–125.

[73] J. Milnor and W. Thurston, On iterated maps of the interval, in: *Dynamical Systems* (J.C. Alexander, ed.), Lect. Notes Math., Vol. 1342, pp. 465–563, Springer, Berlin–Heidelberg, 1988.

[74] C. Mira, L. Gardini, A. Barugola, and J.C. Cathala, *Chaotic dynamics in two-dimensional noninvertible maps*, World Scientific, Singapore, 1996.

[75] M. Misieurewicz, Strange attractors for the Lozi mapping, *Ann. NY Acad. Sci.* **80** (1980), 348–358.

[76] Yu.I. Neimark and S.D. Kinyapin, On the equilibrium state on a surface of discontinuity, *Izv. VUZ. Radiofizika* **3** (1960), 694–705.

[77] G. Nikolsky, On automatic stability of a ship on a given course, *Proc. Central Commun. Lab.* (in Russian) **1** (1934), 34–75.

[78] A.B. Nordmark, Universal limit mapping in grazing bifurcations, *Phys. Rev. E.* **55** (1997), 266–270.

[79] D.N. Novaes and M.R. Jeffrey, Regularization of hidden dynamics in piecewise-smooth flow, *J. Differ. Equ.* **259** (2015), 4615–4633.

[80] H.E. Nusse, E. Ott, and J.A. Yorke, Border-collision bifurcations: an explanation for observed bifurcation phenomena, *Phys. Rev. E* **49** (1994), 1073–1076.

[81] H.E. Nusse and J.A. Yorke, Border-collision bifurcation including 'period two to period three' for piecewise-smooth systems, *Physica D* **57** (1992), 39–57.

[82] J. Palis and W. de Melo, *Geometric theory of dynamical systems*, Springer, 1982.

[83] F. Rhodes and C.L. Thompson, Rotation numbers for monotone functions on the circle, *J. London Math. Soc.* **34** (1986), 360–368.

[84] F. Rhodes and C.L. Thompson, Topologies and rotation numbers for families of monotone functions on the circle, *J. London Math. Soc.* **43** (1991), 156–170.

[85] D.J.W. Simpson, Sequences of periodic solutions and infinitely many coexisting attractors in the border-collision normal form, *Int. J. Bif. & Chaos* **24** (2014), 1430018.

[86] D.J.W. Simpson, On the relative coexistence of fixed points and period-two solutions near border-collision bifurcations, *Appl. Math. Lett.* **38** (2014), 162–167.

[87] D.J.W. Simpson, Border-collision bifurcations in $\mathbb{R}^N$, *SIAM Review* **58** (2016), 175–175.

[88] D.J.W. Simpson and J.D. Weiss, Shrinking point bifurcations of resonance tongues for piecewise-smooth, continuous maps, *Nonlinearity* **22** (2009), 1123–1144.

[89] D.J.W. Simpson and J.D. Meiss, Resonance near border-collision bifurcations in piecewise-smooth, continuous maps, *Nonlinearity* **23** (2010), 3091–3118.

[90] J. Sotomayor and M.A. Teixeira, Regularization of discontinuous vector fields, in: *Proceedings of the International Conference on Differential Equations, Lisboa (1996)*, pp. 207–223.

[91] M.A. Teixeira, Structural stability of pairings of vector fields and functions, *Boletim da S.B.M.* **9** no. 2 (1978), 63–82.

[92] M.A. Teixeira, Stability conditions for discontinuous vector fields, *J. Differ. Equ.* **88** (1990), 15–29.

[93] M.A. Teixeira, Generic bifurcation of sliding vector fields, *J. Math. Anal. Appl.* **176** (1993), 436–457.

[94] M. Tsujii, Absolutely continuous invariant measures for expanding piecewise linear maps, *Invent. Math.* **143** (2001), 349–373.

[95] V.I. Utkin, Variable structure systems with sliding modes, *IEEE Trans. Automat. Contr.* **22** (1977), 212–222.

[96] V.I. Utkin, *Sliding modes and their application in variable structure systems*, vol. 1 (Translated from the Russian), MiR, 1978.

[97] V.I. Utkin, *Sliding modes in control and optimization*, Springer-Verlag, 1992.

[98] V.I. Utkin, *Sliding mode control in electro-mechanical systems*, CRC Press, 1999.

[99] A.J. van der Schaft and J.M. Schumacher, *An introduction to hybrid dynamical systems*, Springer-Verlag, 2000.

[100] R.F. Williams, The structure of Lorenz attractors, *Publ. Math. IHES* **50** (1979), 73–99.

[101] L.S. Young, Bowen–Ruelle measures for certain piecewise hyperbolic maps, *Trans. Amer. Math. Soc.* **287** (1985), 41–48.

[102] X. Zhang, Sinai–Ruelle–Bowen measures for piecewise hyperbolic maps with two directions of instability in three-dimensional spaces, *Discrete Cont. Dyn. Syst.* **36** (2016), 2873–2886.

Printed in the United States
By Bookmasters